Higher Core
Physics

Higher Core Physics

Geoff Cackett
Assistant Head Teacher, Holy Rood High School, Edinburgh

Jim Lowrie
Senior Physics Lecturer, Telford College, Edinburgh

Alastair Steven
Principal Physics Teacher, Wester Hailes Education Centre, Edinburgh

Oxford University Press

Oxford University Press, Walton Street, Oxford OX2 6DP

Oxford   New York   Toronto
Delhi   Bombay   Calcutta   Madras   Karachi
Petaling Jaya   Singapore   Hong Kong   Tokyo
Nairobi   Dar es Salaam   Cape Town
Melbourne   Auckland

and associated companies in
Berlin   Ibadan

Oxford is a trade mark of Oxford University Press

First published 1983
Reprinted 1984, 1985, 1988, 1989

ISBN 0 19 914096 0

Cover illustration by
Colin Rattray.

Phototypeset by Tradespools Limited,
Frome, Somerset

Printed by Butler & Tanner Limited,
Frome, Somerset

Preface

This book is written specifically for students preparing for Scottish Higher Grade Physics and takes full account of the revised syllabus to be first examined in 1984. The text is designed to equip students with the understanding, basic knowledge and problem-solving skills required at this level.

The Higher grade syllabus is an extension of and includes the work of the Ordinary Grade syllabus. This book was written on this basis and follows the same style as *Core Physics* (for Ordinary Grade) in dealing with all the objectives in sections N, O, P, and Q of the revised syllabus: the remaining section R is dealt with in memoranda from the Scottish Curriculum Development Service in Dundee.

Each chapter covers a particular topic of the syllabus with a full explanation. The practical nature of the course is emphasized by reference to many experiments with diagrams and photographs. Sample results are used to derive relationships, and worked examples have been used to illustrate particular points and to help with understanding. Each chapter ends with a summary and problems, many of the problems being from past Higher Grade papers. Numerical answers to all problems have been provided; SI units are used throughout, other units being referred to only when they are in common use. The negative index notation for units has been used (e.g. $m\,s^{-1}$ rather than m/s) in accordance with examination requirements. 'Electron flow' current convention is used and conventional current is not used at all.

We should like to express our appreciation to the various people and establishments listed on page 208 for permission to reproduce drawings and photographs, and to the Scottish Examination Board for allowing us to include questions from past Higher Grade examination papers.

Finally, we should like to thank Ellice, Eleanor and Sylvia for their support and encouragement.

Geoff Cackett
Jim Lowrie
Alastair Steven

Contents

1 Kinetics

1.1 Uniformly accelerated motion in a straight line

When an object moves with uniform acceleration in a straight line it is often necessary to predict some of the quantities involved such as displacement or the velocity attained after a given time. This is most usefully done by developing a set of equations usually known as the **equations of motion.** It is essential to remember that these are valid for **uniform** acceleration in a straight line only and cannot be used when the acceleration is variable.

The following symbols will be used:

u initial velocity	s displacement	t time interval
v final velocity	a uniform acceleration	

We shall consider motion in a straight line only, so that the vector quantities velocity and displacement will not have a direction quoted. When an object such as a car is moving, the changes in motion which take place over a given period of time can be displayed in the form of a graph. For example a car moving along a motorway at a steady velocity of $30\,\mathrm{m\,s^{-1}}$ (70 mph) will have the velocity-time graph shown in Figure 1.1. Because the velocity is uniform, the displacement in each second will be the same, namely 30 metres, so that after one second the displacement will be $30\,\mathrm{m}$ and after two seconds $60\,\mathrm{m}$. This is shown graphically in Figure 1.2. After t seconds the displacement will be $30 \times t$ metres.

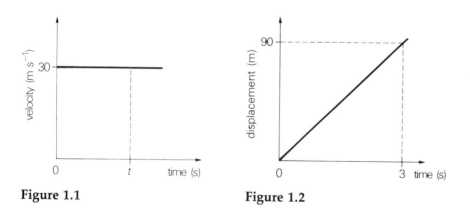

Figure 1.1   **Figure 1.2**   **Figure 1.3**

If however the car starts from rest and accelerates at $2\,\mathrm{m\,s^{-2}}$, the velocity will be changing by $2\,\mathrm{m\,s^{-1}}$ every second and the displacements in successive seconds will be continuously increasing.

A velocity-time graph of this motion, Figure 1.3, can be constructed as follows.

At the start the velocity will be zero

After 1 second the velocity will be $2\,\mathrm{m\,s^{-1}}$

After 2 seconds the velocity will be $4\,\mathrm{m\,s^{-1}}$

After 3 seconds the velocity will be $6\,\mathrm{m\,s^{-1}}$

After 4 seconds the velocity will be $8\,\mathrm{m\,s^{-1}}$

1

The acceleration can be found by calculating the gradient of the velocity time graph. In Figure 1.4, the velocity is u at time t_1 and v at time t_2.

$$\text{acceleration} = \frac{\text{change of velocity}}{\text{time interval}}$$

$$a = \frac{v - u}{t_2 - t_1} = \text{gradient of graph}$$

using values from Figure 1.3,

$v = 8\,\text{m s}^{-1}$ when $t_2 = 4\,\text{s}$

$u = 2\,\text{m s}^{-1}$ when $t_1 = 1\,\text{s}$

$\Rightarrow \qquad a = \dfrac{8 - 2}{4 - 1} = \dfrac{6}{3} = 2$

The acceleration is $2\,\text{m s}^{-2}$.

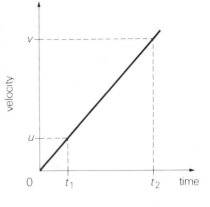

Figure 1.4

1.2 Displacement-time graph

The displacement-time graph associated with the motion represented in Figure 1.5 can be constructed in the following way.

For uniform acceleration

the average velocity is given by $\bar{v} = \dfrac{u + v}{2}$

and the displacement is given by $s = \text{average velocity} \times \text{time}$.

We can apply this to the data given in Figure 1.5.

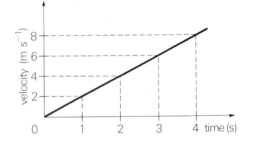

displacement in first second

   initial velocity   = 0; final velocity = $2\,\text{m s}^{-1}$

   average velocity = $1\,\text{m s}^{-1}$

   displacement   = $1\,\text{m}$

Figure 1.5

displacement in second second

   initial velocity   = $2\,\text{m s}^{-1}$; final velocity = $4\,\text{m s}^{-1}$

   average velocity = $3\,\text{m s}^{-1}$

   displacement   = $3\,\text{m}$

   the total displacement after 2 seconds = $1 + 3 = 4\,\text{m}$

displacement in third second

   initial velocity   = $4\,\text{m s}^{-1}$; final velocity = $6\,\text{m s}^{-1}$

   average velocity = $5\,\text{m s}^{-1}$

   displacement   = $5\,\text{m}$

   the total displacement after 3 seconds = $4 + 5 = 9\,\text{m}$

displacement in fourth second

   initial velocity   = $6\,\text{m s}^{-1}$; final velocity = $8\,\text{m s}^{-1}$

   average velocity = $7\,\text{m s}^{-1}$

   displacement   = $7\,\text{m}$

   the total displacement from the start = $9 + 7 = 16\,\text{m}$

The displacement-time graph is shown in Figure 1.6.

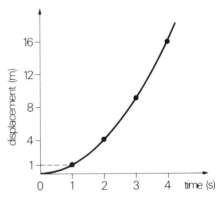

Figure 1.6

1.3 Equations of motion

Using graphical methods, we can now develop equations which can be used to analyse uniformly accelerated motion.

Figure 1.7 shows the velocity–time graph for a body which, initially moving at velocity u accelerates at a for time t until the final velocity is v

final velocity = initial velocity + increase in velocity but after a time t the increase will be at. The equation will be

$$v = u + at \quad \text{...[1]}$$

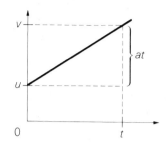

Figure 1.7

If the same graph is redrawn as in Figure 1.8, it can be seen that the area under the graph can be divided into a rectangle and a triangle. The total area gives the displacement.

For the rectangle: base $= t - 0 = t$; height $= u - 0 = u$

$\Rightarrow$ area $= ut$

For the triangle: base $= t - 0 = t$; height $= v - u$,

but from equation [1], $v - u = at$

$\Rightarrow$ area $= \frac{1}{2} \times$ base $\times$ height

$= \frac{1}{2} \times t \times at$

$= \frac{1}{2}at^2$

$\therefore$ total displacement = area of rectangle + area of triangle

$\Rightarrow$ $s = ut + \frac{1}{2}at^2 \quad \text{...[2]}$

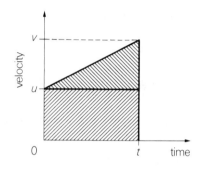

Figure 1.8

By combining equations [1] and [2], a third equation can be developed which does not involve time.

From equation [1], $t = \dfrac{v - u}{a}$

and substituting in equation [2]

$$s = u\left(\frac{v - u}{a}\right) + \frac{1}{2}a\left(\frac{v - u}{a}\right)^2$$

$$= \frac{uv - u^2}{a} + \frac{v^2 - 2uv + u^2}{2a}$$

$\Rightarrow$ $2as = 2uv - 2u^2 + v^2 - 2uv + u^2$

$= v^2 - u^2$

$\Rightarrow$ $v^2 = u^2 + 2as \quad \text{...[3]}$

The acceleration term can be eliminated between equations [1] and [2] to give equation [4].

From equation [1], $a = \dfrac{v - u}{t}$

and substituting in equation [2]

$$s = ut + \frac{1}{2}at^2$$

$$= ut + \frac{1}{2}\left(\frac{v - u}{t}\right)t^2$$

$$= ut + \frac{1}{2}vt - \frac{1}{2}ut$$

$$= \frac{1}{2}ut + \frac{1}{2}vt$$

$\Rightarrow$ $s = \left(\dfrac{u + v}{2}\right)t \quad \text{...[4]}$

The term $\left(\dfrac{u + v}{2}\right)$ is the average velocity $\bar{v}$ if acceleration is uniform.

Example 1

A car starts from rest and accelerates uniformly at $3\,\mathrm{m\,s^{-2}}$.

a) How long will it take to reach a velocity of $30\,\mathrm{m\,s^{-1}}$?
b) What distance will it have travelled in this time?

a) $u = 0$, $a = 3\,\mathrm{m\,s^{-2}}$, $v = 30\,\mathrm{m\,s^{-1}}$, $t = ?$

Using equation [1], $v = u + at$

$$\Rightarrow \qquad 30 = 0 + 3t$$

$$\Rightarrow \qquad t = \tfrac{30}{3} = 10$$

The car takes 10 seconds to reach a velocity of $30\,\mathrm{m\,s^{-1}}$.

b) $u = 0$, $v = 30\,\mathrm{m\,s^{-1}}$, $t = 10\,\mathrm{s}$, $s = ?$

Using equation [4], $\qquad s = \left(\dfrac{u+v}{2}\right)t$

$$\Rightarrow \qquad s = \left(\dfrac{0+30}{2}\right)10$$

$$= 15 \times 10 = 150$$

The distance travelled is 150 m.

1.4 Acceleration produced by gravity

The acceleration g produced by the gravitational field of the Earth is always directed towards the centre of the Earth.

It is usual to measure displacement in a direction away from the centre of the Earth so that all vectors upwards are positive and all vectors downwards are negative.

In this book, the approximate value of acceleration due to gravity is always taken to be $-10\,\mathrm{m\,s^{-2}}$.

positive vectors	negative vectors
upward displacement	downward displacement
upward velocity	downward velocity
	gravitational acceleration

Example 2

If a ball is dropped from a window, what is its velocity 3 seconds later?

$u = 0$, $a = -10\,\mathrm{m\,s^{-2}}$, $t = 3\,\mathrm{s}$, $v = ?$

Using equation [1], $v = u + at$

$$\Rightarrow \qquad v = 0 + (-10) \times 3 = -30$$

The velocity of the ball 3 seconds later is $30\,\mathrm{m\,s^{-1}}$ downwards.
(The negative sign shows that the velocity of the ball is downwards)

Example 3

An arrow is shot vertically upwards with a velocity of $20\,\mathrm{m\,s^{-1}}$.
a) How long will the arrow take to reach its maximum height?
b) What is the maximum height reached by the arrow?

a) The maximum height is reached when the velocity of the arrow is zero: the velocity then becomes negative as the arrow falls.

$$u = +20\,\mathrm{m\,s^{-1}},\ v = 0\,\mathrm{m\,s^{-1}},\ a = -10\,\mathrm{m\,s^{-2}},\ t = ?$$

Using equation [1], $v = u + at$

$$\Rightarrow \qquad 0 = 20 - 10 \times t$$

$$\Rightarrow \qquad t = \frac{-20}{-10} = 2$$

The arrow takes 2 seconds to reach its maximum height.

b) $u = +20\,\mathrm{m\,s^{-1}},\ v = 0\,\mathrm{m\,s^{-1}},\ t = 2\,\mathrm{s}$

Using equation [4], $s = \left(\dfrac{u+v}{2}\right)t$

$$\Rightarrow \qquad s = \left(\frac{20+0}{2}\right) \times 2$$

$$= 20$$

The maximum height reached is 20 metres.

Example 4

A helicopter is climbing vertically with a velocity of $15\,\mathrm{m\,s^{-1}}$ when an object is released from it. If the object hits the ground 4 s later find

a) the velocity of the object just as it hits the ground
b) the original height of the object.

a) At the instant of release, the object has the same velocity as the helicopter, so that it is moving upwards at $15\,\mathrm{m\,s^{-1}}$. The path of the object is therefore as shown.

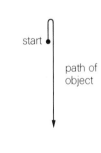

When the object is released: $u = +15\,\mathrm{m\,s^{-1}},\ a = -10\,\mathrm{m\,s^{-2}},\ t = 4\,\mathrm{s},\ v = ?$

Using equation [1], $v = u + at$

$$\Rightarrow \qquad v = +15 - 10 \times 4$$

$$= +15 - 40 = -25$$

The velocity of the object just as it hits the ground is $25\,\mathrm{m\,s^{-1}}$ downwards.

b) To calculate the original height of the object, we must find the displacement of the object from the start.

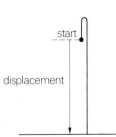

Using equation [2], $s = ut + \frac{1}{2}at^2$

$$\Rightarrow \qquad s = +15 \times 4 - \frac{1}{2} \times 10 \times 4^2$$

$$= +60 - 80 = -20$$

The displacement from the starting point is 20 m downwards so that the original height is 20 m.

Example 5

An object is projected vertically upwards with a velocity of $40\,\mathrm{m\,s^{-1}}$
a) Find the time taken to reach the maximum height.
b) Calculate the maximum height reached.
c) Find the time taken to fall back to the starting point.
d) Draw a graph showing the variation of velocity with time until the object hits the ground.

a) At the maximum height the object stops rising so that the velocity is zero.
$$u = +40\,\mathrm{m\,s^{-1}},\ v = 0,\ a = -10\,\mathrm{m\,s^{-2}}.$$
Using equation [1], $v = u + at$
$$\Rightarrow \qquad 0 = +40 - 10 \times t$$
$$\Rightarrow \qquad -40 = -10t$$
$$\Rightarrow \qquad t = \frac{-40}{-10} = 4$$

The time taken to reach the maximum height is 4 seconds.

b) The maximum height is reached after $4\,\mathrm{s}$
$$t = 4\,\mathrm{s},\ u = +40\,\mathrm{m\,s^{-1}},\ a = -10\,\mathrm{m\,s^{-2}},\ s = ?$$
using $s = ut + \frac{1}{2}at^2$
$$s = 40 \times 4 - \frac{1}{2} \times 10 \times 4^2$$
$$\Rightarrow \qquad s = 160 - 80$$
$$\Rightarrow \qquad s = 80$$
The maximum height is 80 m.

c) When the object has returned to the ground the displacement is zero.
$$s = 0,\ u = +40\,\mathrm{m\,s^{-1}},\ a = -10\,\mathrm{m\,s^{-2}},\ t = ?$$
using $s = ut + \frac{1}{2}at^2$
$$0 = +40 \times t - \frac{1}{2} \times 10 \times t^2$$
$$\Rightarrow \qquad 0 = 40t - 5t^2$$
$$5t^2 - 40t = 0$$
$$\Rightarrow \quad t^2 - 8t = 0$$
$$\Rightarrow \quad t(t - 8) = 0$$
$$t = 0 \text{ or } 8$$
The time taken to return to the starting point is 8 s.

d) In order to construct the velocity-time graph, equation [1] must be applied,
$v = u + at$
After 1 second, $v = 40 - 10 \times 1 = 40 - 10 = 30$
After 2 seconds, $v = 40 - 10 \times 2 = 40 - 20 = 20$
After 3 seconds, $v = 40 - 10 \times 3 = 40 - 30 = 10$
After 4 seconds, $v = 40 - 10 \times 4 = 40 - 40 = 0$

Similarly it is found that
After 6 seconds $v = 40 - 10 \times 6 = 40 - 60 = -20$
After 8 seconds $v = 40 - 10 \times 8 = 40 - 80 = -40$
The graph will therefore be as shown on the right.

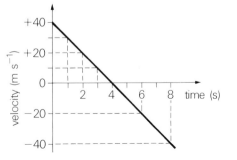

1.5 Measuring acceleration

Acceleration can be measured from a multiflash photograph of a moving object. Figure 1.9 shows a multiflash photograph of a white marker attached to an accelerating vehicle running on a sloping linear air track. A scale showing in centimetres the actual distance travelled is given underneath. The stroboscopic light flashes every 0.1 s. The vehicle started from rest.

Figure 1.9

The spaces between each image are increasing, indicating that the velocity is also increasing. The average velocity during each time interval is given by

$$\text{average velocity} = \frac{\text{displacement during time interval}}{\text{time interval}}$$

In order to calculate the acceleration a table is drawn up.

distance travelled in each 0.1 s interval (m)	0.01	0.03	0.05	0.07	0.09	0.11
average velocity in each 0.1 s interval (m s^{-1})	0.10	0.30	0.50	0.70	0.90	1.10
increase in velocity during each 0.1 s interval (m s^{-1})		0.20	0.20	0.20	0.20	0.20
increase in velocity during a 1 s interval (m s^{-1})		2.00	2.00	2.00	2.00	2.00

Table 1

The acceleration is thus uniform and equal to 2 m s^{-2}

If the acceleration is uniform, its value can be checked from equation [2]:

$$s = ut + \tfrac{1}{2}at^2$$

where s = total displacement from the start.

From Figure 1.9 it can be seen that when $t = 0.6$ s, $s = 0.36$ m,

so that $s = 0.36$, $u = 0$, $t = 0.6$ s, $a = ?$

$$0.36 = 0 \times 0.6 + \tfrac{1}{2}a \times 0.6^2 = 0 + \tfrac{1}{2} \times 0.36 \times a = 0.18a$$

$$\Rightarrow \qquad a = \frac{0.36}{0.18} = 2$$

The acceleration is thus 2 m s^{-2} confirming the result obtained from Table 1.

1.6 Projectile motion

The equations of motion can be used to analyse the motion of a projectile. The simplest case involves a projectile which is fired or thrown horizontally, for example, a bomb leaving an aircraft which is flying in a horizontal direction, Figure 1.10. The path taken by the bomb is curved.

Figure 1.10

Figure 1.11 shows two steel balls released simultaneously, one allowed to fall, the other fired in a horizontal direction.

Notice that both balls fall through the same vertical distance for each time interval. Notice also that the horizontal distance for the ball following the curved path is the same for each time interval because each image is equally spaced horizontally.

We can say that the velocity v of each ball has a vertical component v_y and a horizontal component v_x.

Both balls have the same vertical velocity v_y but for the ball that is falling, the horizontal component v_x is zero.

The ball that is fired horizontally has a constant horizontal velocity v_x because the images are equally spaced horizontally.

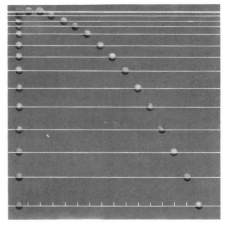

Figure 1.11

Example 6

A ball rolls along a horizontal bench and falls off the end. Figure 1.12 shows the horizontal and vertical distance travelled at intervals of 0.1 s after it has left the bench.

a) Calculate the horizontal component v_x of its velocity.
b) Calculate the vertical component a_y of its acceleration.
c) Find its true velocity v at a time 0.3 s after it has left the bench.

time (s)	0.1	0.2	0.3	0.4
horizontal distance (m)	0.15	0.30	0.45	0.60
vertical distance (m)	0.05	0.20	0.45	0.80

a) For each time interval of 0.1 s the ball travels a constant horizontal distance of 0.15 m; the horizontal component v_x of its velocity is constant.

$$v_x = \frac{\text{horizontal distance}}{\text{time interval}}$$

$$= \frac{0.15}{0.1} = 1.5$$

This is the velocity of projection from the horizontal bench.

The horizontal component of velocity (velocity of projection) is 1.5 m s^{-1}.

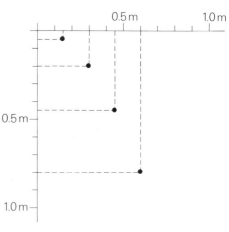

b)

time interval (s)	0.1	0.1	0.1	0.1
vertical distance interval (m)	0.05	0.15	0.25	0.35
average vertical velocity (m s^{-1})	0.5	1.5	2.5	3.5

Between each image the average vertical velocity increases as follows:

$(1.5 - 0.5) = 1.0 \text{ m s}^{-1}$ in 0.1 s

$(2.5 - 1.5) = 1.0 \text{ m s}^{-1}$ in 0.1 s

$(3.5 - 2.5) = 1.0 \text{ m s}^{-1}$ in 0.1 s

So the acceleration a_y is constant:

$$a_y = \frac{\text{vertical velocity interval}}{\text{time interval}} = \frac{1.0}{0.1} = 10$$

The vertical component of acceleration is 10 m s^{-2}.

Figure 1.12

c) The actual velocity at any instant is found by combining the horizontal velocity component v_x with the vertical velocity component v_y at that instant in a triangle of velocities.

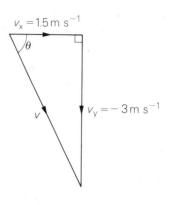

v_x is constant and equals $1.5\,\mathrm{m\,s^{-1}}$

The value of v_y when $t = 0.3$ is found from equation [1],

$v_y = u_y + a_y t$ where $a = -10\,\mathrm{m\,s^{-2}}$ and $u_y = 0\,\mathrm{m\,s^{-1}}$

$\Rightarrow v_y = 0 - 10 \times 0.3$

$\qquad = -3$

The triangle of velocities shown is used to obtain the velocity v from v_x and v_y after $0.3\,\mathrm{s}$.

By Pythagoras' theorem $v_2 = v_x^2 + v_y^2$

$\qquad\qquad\qquad\qquad = 1.5^2 + (-3)^2$

$\qquad\qquad\qquad\qquad = 11.25$

$\Rightarrow\qquad\qquad\qquad v = 3.35$

The angle θ that the actual velocity makes with the horizontal is given by

$\tan\theta = \dfrac{v_y}{v_x} = \dfrac{-3}{1.5} = -2.0$

$\Rightarrow\quad \theta = -63.4°$

The velocity after $0.3\,\mathrm{s}$ is $3.35\,\mathrm{m\,s^{-1}}$ at an angle of $63.4°$ below the horizontal.

Example 7

A ball is projected horizontally off the end of a bench. It hits the ground $3\,\mathrm{m}$ from the base of the bench and a vertical distance $1.25\,\mathrm{m}$ below the point of projection. Find **a)** the time of flight and **b)** the velocity of projection.

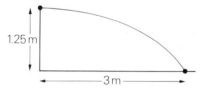

a) Consider the vertical motion. The distance s_y is negative because it is downwards.

$s_y = -1.25\,\mathrm{m}$, $u_y = 0$, $a_y = -10\,\mathrm{m\,s^{-2}}$

Using the equation $s = ut + \frac{1}{2}at^2$

$\Rightarrow\qquad -1.25 = 0 + \frac{1}{2} \times -10 \times t^2$

$\Rightarrow\qquad t^2 = \dfrac{-1.25}{-5} = 0.25$

$\Rightarrow\qquad t = \pm 0.5$ (the negative value has no meaning)

The time of flight is $0.5\,\mathrm{s}$.

b) Consider the horizontal motion.

$\text{velocity } v_x = \dfrac{\text{horizontal displacement}}{\text{time interval}}$

$\qquad\qquad = \dfrac{3}{0.5} = 6$

The velocity of projection is $6\,\mathrm{m\,s^{-1}}$.

When a golfer drives the ball down the fairway, the ball is projected at an angle to the horizontal, Figure 1.13. We can see that the ball has both a horizontal component and an upwards (positive) vertical component when it is hit. So that u_y is not zero.

Figure 1.13

If air resistance is neglected, the path of the ball is symmetrical about the point of maximum height as shown in Figure 1.14. A similar motion is given in Figure 1.15. The range R is the total horizontal distance travelled, AQ in Figure 1.14.

The maximum height is OP.

The time of flight is the time taken to return to the horizontal again, that is to go from point A to point Q.

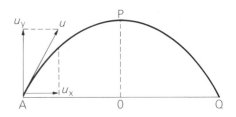

Figure 1.14

Example 8

A golfer hits a ball with a velocity of $48\,\mathrm{m\,s^{-1}}$ at an angle of $30°$ to the horizontal. If air resistance is neglected, find **a)** the time of flight, **b)** the range, **c)** the maximum height

a) The time of flight is the time taken for the ball to return to the ground (assumed horizontal), i.e. $s_y = 0$

The vertical component u_y of the velocity is found from the triangle ADC.

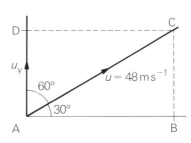

$$\frac{AD}{AC} = \frac{u_y}{u} = \cos 60°$$

$$\Rightarrow \quad u_y = u \cos 60°$$

$$= 48 \times 0.5 = 24$$

Using the equation $s_y = u_y t + \frac{1}{2}a_y t^2$
where $s_y = 0$, $u_y = +24\,\mathrm{m\,s^{-1}}$, $a_y = -10\,\mathrm{m\,s^{-2}}$, $t = ?$

$$\Rightarrow 0 = 24t + \frac{1}{2} \times -10 \times t^2$$

$$\Rightarrow 0 = t(24 - 5t)$$

$$\Rightarrow t = 0 \text{ or } \frac{24}{5}$$

$$= 0 \text{ or } 4.8$$

The time of flight is 4.8 seconds.

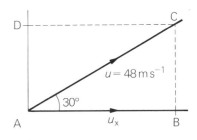

b) The range is the total horizontal displacement s_x for the time of flight and is found from the horizontal component u_x of the velocity.

From triangle ABC

$$\frac{AB}{AC} = \frac{u_x}{u} = \cos 30°$$

$$\Rightarrow \quad u_x = u \cos 30°$$

$$= 48 \times 0.87 = 41.8$$

$$\Rightarrow \quad s_x = u_x t = 41.8 \times 4.8 = 200$$

The range is 200 metres.

c) The maximum height is gained when the vertical component v_y of the velocity is zero.

Using the equation $v_y^2 = u_y^2 + 2a_y s_y$

$$v_y = 0, \ u_y = +24\,\mathrm{m\,s^{-1}}, \ a_y = -10\,\mathrm{m\,s^{-2}}, \ s_y = ?$$

$$\Rightarrow \qquad 0 = 24^2 + 2 \times -10\,s_y$$

$$\Rightarrow \qquad s_y = \frac{576}{20} = 28.8$$

The maximum height is 28.8 metres.

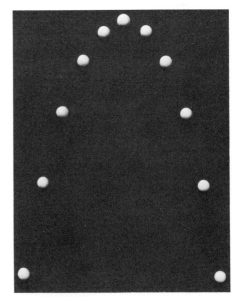

Figure 1.15

Summary

Acceleration is given by the gradient of the velocity-time graph

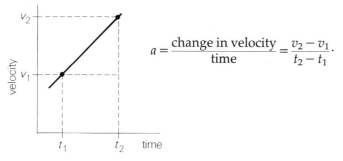

$$a = \frac{\text{change in velocity}}{\text{time}} = \frac{v_2 - v_1}{t_2 - t_1}.$$

The distance travelled by a moving object is equal to the area under the velocity-time graph.

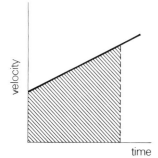

The equations of motion are

$$v = u + at$$
$$s = ut + \tfrac{1}{2}at^2$$
$$v^2 = u^2 + 2as$$
$$s = \left(\frac{u + v}{2}\right)t$$

The instantaneous velocity of a projectile is the vector sum of the horizontal and vertical components of velocity.

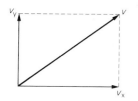

Problems

1   a) What is meant by a scalar quantity?
b) Classify the following as either vector or scalar quantities: speed, displacement, velocity, acceleration, distance
c) One of the equations of motion is $s = ut + \tfrac{1}{2}at^2$. Write down each symbol and state whether it is a vector or scalar.

2   A person walks due North at $1.5\,\text{m s}^{-1}$ across the deck of a ship which is travelling due West at $2\,\text{m s}^{-1}$. Determine the magnitude and direction of the velocity of the person relative to the water.

3   A fork-lift truck raises a box vertically, then moves 3.5 m across the floor of a warehouse and deposits the box on a platform 1.5 m high. Calculate the displacement of the box from the starting position.

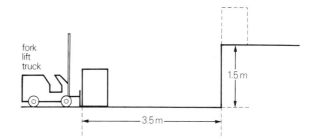

4   a) A train enters a tunnel which is 180 m long and emerges 6 s later. What is the average speed of the train as it goes through the tunnel?
b) An electron takes 1.2 nanoseconds to travel a distance of 10 cm. Calculate the speed of the electron.
c) A runner travels a distance of 1500 m in a time of 3 minutes 55 seconds. Calculate his average speed.

5   A reproduction of a multi-flash photograph is shown. The flash-rate is 8 Hz.

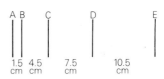

a) Calculate the average speeds between AB and DE and hence find the acceleration.
b) What is the average speed for the motion between A and E?

6   An object slides from rest down a slope and the positions at intervals of 0.2 s are shown.

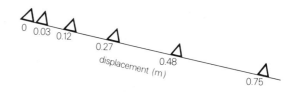

a) Calculate the acceleration.
b) Determine the displacement of the object after a further 0.2 s.

7 a) Describe an experiment to find the acceleration of a moving trolley using a ticker-timer and tape. State what measurements would be taken and show how they would be used to calculate the acceleration.
b) Explain how you would check that the reading of 48 km/h on a car speedometer is correct.
c) How could the frequency of rotation of the shaft of an electric motor be determined using a multi-flash stroboscope?
d) You are asked to determine the acceleration of gravity g using a multi-flash photograph of a falling golf ball. Describe how you would do this, clearly explaining the measurements you would take and how you would calculate a value for g.

8 A ball is thrown vertically upwards with a velocity of $15\,\text{m s}^{-1}$.
a) How long will it take to fall back to the starting position?
b) What distance has the ball travelled in this time?

9 A coin rests on a record-player turntable a distance of 10 cm from the spindle. When the turntable is revolving at 45 r.p.m., what is the average speed of the coin? Is the velocity of the coin changing?

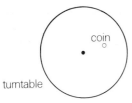
coin
turntable

10 Describe three situations in which an object moves with varying acceleration.

11 A vehicle is catapulted along a horizontal linear air track. The friction force acting on the vehicle is negligible.

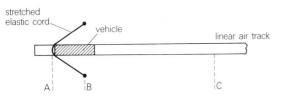
stretched elastic cord
vehicle
linear air track
A   B   C

Describe the acceleration of the vehicle during sections AB and BC.

12 The speed-time graph shows how the speed of a trolley varies as it runs down a slope. How far did the trolley travel in the first 3 seconds?

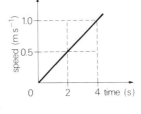

13 The speed-time graph for a runner is shown.
a) Construct an acceleration-time graph.
b) How far did she run?

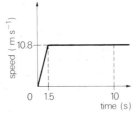

14 The velocity-time graph for a moving object is shown. What is the displacement after 6 s?

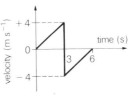

15 The total displacements of an object moving in a straight line at given times are listed in the table.

displacement s (m)	0	2	8	18	32	50
time t (s)	0	1	2	3	4	5

a) Is the object accelerating or decelerating?
b) Construct a speed-time graph for the motion.

16 The acceleration of an object varies with time as shown. If the initial speed is zero construct a speed-time graph.

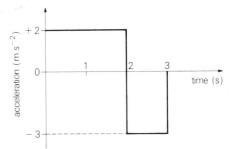

a) Determine the speed after 3 seconds.
b) How far has the object travelled in this time?

17 The displacement-time graphs for the motion of three objects is shown. In each case calculate the acceleration.

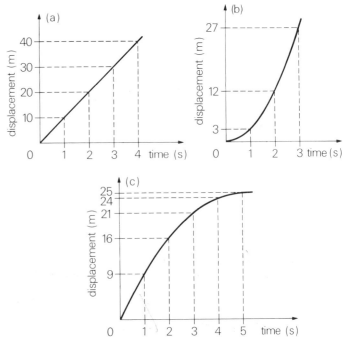

18 The acceleration of a car varies as shown.

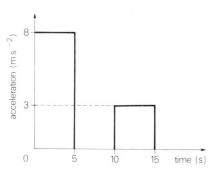

If the initial velocity is zero, draw a speed-time graph. How far did the car travel during the period 10 s to 15 s?

19 A helicopter is rising vertically at 8 m s^{-1}. An object is released from it and falls. The object hits the ground 3 s later.
a) What is the velocity of the object as it hits the ground?
b) Determine the height of the object when it was released.

20 For each of the following situations, sketch possible velocity-time graphs. A ball is
a) thrown vertically upwards and caught when it returns to the starting point.
b) dropped to the floor and caught after the first bounce.
c) rolled along a horizontal surface and then accelerated down a sloping ramp.
d) rolled along a horizontal surface, up an incline, and then back down to stop finally on the horizontal surface.

21 An object is accelerated uniformly from rest. The displacement-time graph is shown.

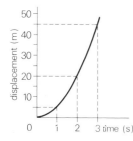

Giving actual values, draw
a) the corresponding velocity-time graph
b) the corresponding acceleration-time graph.

22 A ball rolls over the edge of a horizontal surface as shown.

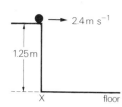

a) How long will it take to hit the floor?
b) Calculate the vertical component of the velocity as it hits the floor.
c) If the horizontal velocity of the ball is 2.4 m s^{-1}, determine the angle at which the ball hits the floor.
d) How far from point X did the ball land?

23 A rolling ball leaves the end of a sloping ramp at 5 m s^{-1}, as shown in the diagram.
a) How long will it take to reach the ground?
b) How far will it have travelled horizontally before hitting the ground?

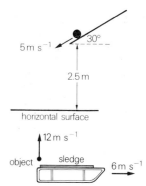

24 A sledge is moving at 6 m s^{-1} along a horizontal surface. An object is projected vertically upwards from it at 12 m s^{-1}.

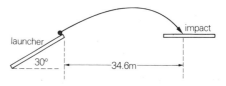

a) Calculate the maximum height reached by the object.
b) Calculate the horizontal distance travelled by the object before it falls back to the same height as the top of the sledge.
c) Sketch the path taken by the object as seen by a stationary observer.
d) What impression of the motion would someone on the sledge have?

25 A metal ball is launched at an angle of 30° and hits a horizontal surface at the same height as the top of the launcher. The point of impact is 34.6 m away and the ball takes 2 seconds to travel this distance.

Find
a) the horizontal component of the velocity of the ball
b) the maximum height reached.

26 A projectile is fired with velocity v from A at an angle α to a horizontal site. It returns to the ground at R.

The horizontal and vertical components of its velocity for the flight are shown in graphs I and II respectively.

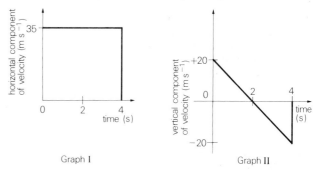

Graph I   Graph II

a) How far from A is the projectile when it hits the ground at R?
b) There is a disused building 15 m high, midway between A and R. What is the distance between the top of this building and the projectile as it passes directly over the building?
c) Calculate the initial speed and direction of the projectile.

SCEEB

1 Kinetics

27 In an experiment to investigate how different factors affect the stopping of an unladen van, a series of speed-time graphs is obtained. In these graphs the speed is recorded from the instant the driver is asked to apply the brakes. One such graph is shown.

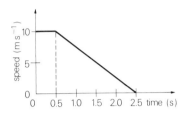

The following are deduced from the series of graphs.

Initial speed of van	Thinking distance	Braking distance	Overall stopping distance
$(m\,s^{-1})$	(m)	(m)	(m)
10	5	10	15
20	10	40	50
30	15	90	105

a) i) What is meant by the term 'thinking distance'?
ii) Explain why thinking distance varies directly as the van's speed.
b) Using the information from the table, reproduce the speed-time graph for the case in which the initial speed of the van is $30\,m\,s^{-1}$.

SCEEB

28 In a game, three boys, Bill, John and Peter, started at the same place P and each ran to one of the three positions marked A, B and C. The distances are marked.

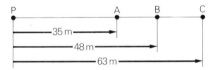

The graph of velocity against time from the start is shown for each boy.

a) Which position did Bill arrive at and what was his average speed for the journey?

b) Copy and complete the following table for the first four seconds of John's journey and use the results to plot a graph of his displacement against time for this period.

Time (s)	0	1	2	3	4
Displacement from start (m)					

c) i) Where was Peter and what was he doing 8 seconds after starting?
ii) What was he doing 10 seconds after starting?

SCEEB

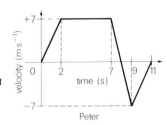

29 The diagram represents the apparatus set up to determine a value for g, the acceleration due to gravity. A steel ball B is held by an electromagnet E at a height h vertically above switch S_2. When switch S_1 is opened, the electric clock is started and E is disconnected from the electrical supply. The ball falls and strikes S_2 which opens and stops the clock.

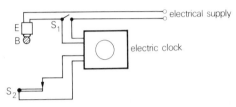

a) Show that the time t for a ball to fall freely from rest through a height h is given by $t = \sqrt{(2h/g)}$.
b) It is found that when $h = 0.600\,m$, the recorded time of fall is $0.400\,s$. Using the above relationship what value do these results give for g?
c) The class decides to repeat the experiment for different values of h, and from the results the graph below is drawn of t against $\sqrt{h}$. Time t is in seconds and h is in metres.

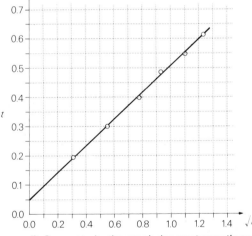

i) Suggest why the graph does not pass through the origin.
ii) From the graph of t against $\sqrt{h}$ find the true time for the ball to fall through a height of $1.44\,m$.
iii) What value does this now give for g?

SCEEB

30 A ball is projected at $15\,m\,s^{-1}$ horizontally from the top of a vertical cliff and reaches the horizontal ground $45\,m$ from the foot of the cliff.
a) Draw accurate graphs, with the appropriate numerical scales, of
i) the horizontal speed of the ball against time
ii) the vertical speed of the ball against time
for the period from its projection until it hits the ground.
b) By using a vector diagram or otherwise, find the velocity of the ball 2 s after its projection giving both speed and direction. State any assumptions you have made.

SCEEB

2 Forces

2.1 Introduction

When a force acts on an object, it can produce a change in velocity either by changing its speed or by changing its direction. A force can also produce a change of shape. A change of velocity will only take place if the forces acting on the object fail to balance out. When the forces do balance, the object does not change its velocity. If the object is at rest, it remains at rest. If the object is moving, it continues to do so with the same speed and direction.

Figure 2.1 shows a multiflash photograph of a straw marker attached to a vehicle moving along a horizontal linear air track. In this case friction is negligible so that the net horizontal force is virtually zero. The distances between the images of the marker are equal, indicating a constant horizontal velocity.

Figure 2.1

An object can be moving at a steady speed but can change direction because of a resultant force. This is illustrated in Figure 2.2 which shows a multiflash photograph of a frictionless puck being whirled at constant speed in a circle. The attached string exerts a force which continually changes the direction of travel of the puck so that it travels in a circle rather than in a straight line.

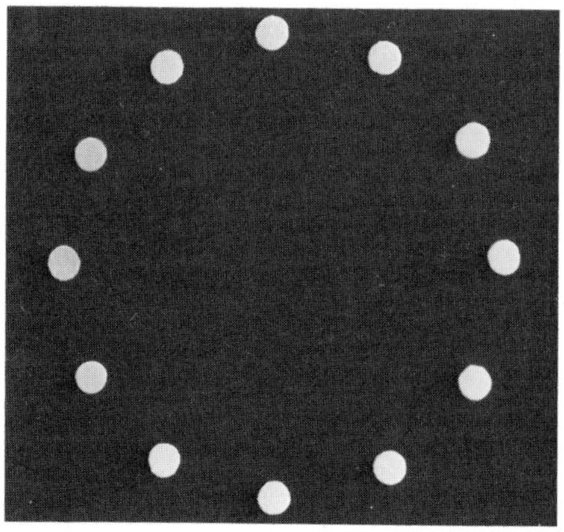

Figure 2.2. 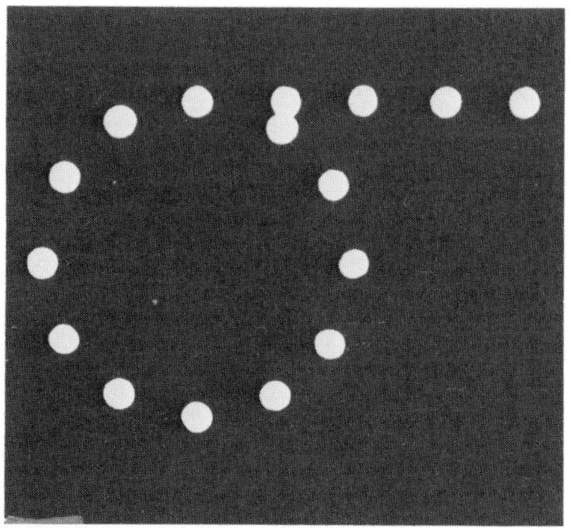   **Figure 2.3**

If the string breaks (Figure 2.3) the puck shoots off at a tangent to the circular path. The equal spacing between the images indicates constant speed. Newton summarised these observations in his First Law which can be formally stated

> 'An object will remain at rest or will continue to move in a straight line at constant speed unless it is acted upon by a net or unbalanced force.'

Every object near the surface of the earth accelerates downwards. An example of this is shown in Figure 2.4.

The spacing between the images increases showing that the ball is accelerating. This acceleration is caused by the unbalanced force acting on the ball.

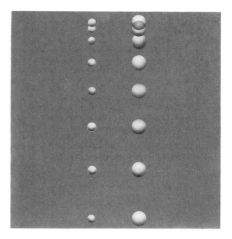

Figure 2.4

The relationship between an unbalanced force F acting on mass m producing acceleration a is given by Newton's Second Law:

$$F = ma$$

Example 1

A trolley with a marker attached is pulled by a stretched elastic cord the extension of which is kept constant. A multiflash photograph is taken at a flash rate of 10 Hz.

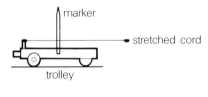

If the mass of the trolley is 0.8 kg, find the unbalanced force acting on the trolley. The distances travelled by the marker in the time interval of 0.1 s are shown in the diagram.

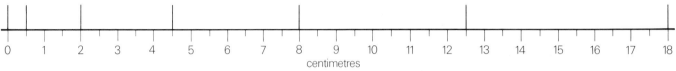

centimetres

The acceleration can be found by constructing a table.

Distance moved in successive 0.1 s intervals	0.005 m	0.015 m	0.025 m	0.035 m	0.045 m	0.055 m
Time interval	0.1 s	0.1 s	0.1 s	0.1 s	0.1 s	0.1 s
Average speed in each 0.1 s interval (m s^{-1})	0.05	0.15	0.25	0.35	0.45	0.55
Increase in speed between successsive 0.1 s intervals (m s^{-1})		0.10	0.10	0.10	0.10	0.10
Increase in speed between successive 1 s intervals (m s^{-1})		1.00	1.00	1.00	1.00	1.00
Acceleration (m s^{-2})		1.00	1.00	1.00	1.00	1.00

Table 1

From the table we can see that the acceleration is uniform and equal to 1 m s^{-2}. The unbalanced force F acting on the trolley is given by

$$F = ma$$
$$= 0.8 \times 1$$
$$= 0.8$$

The unbalanced force acting on the trolley is 0.8 N.

Example 2

An unbalanced force of 25 N acts on a mass of 5 kg. What is the acceleration produced?

Using $F = ma$

$$25 = 5 \times a$$
$$\Rightarrow a = \frac{25}{5}$$
$$= 5$$

The acceleration produced is 5 m s^{-2}.

Example 3

A net force of 24 N acts on two blocks A and B.

a) What is the acceleration of each block?
b) What is the net force acting on block A?

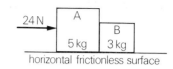

horizontal frictionless surface

a) The blocks will move together so that

total mass $= 5 + 3 = 8$ kg

the acceleration a is given by

$$a = \frac{F}{m}$$
$$= \frac{24}{8} = 3$$

The acceleration of each block is 3 m s^{-2}.

b) The net force acting on block A

$=$ mass of block A $\times$ acceleration of block A

$= 5 \times 3 = 15$

The net force acting on block A is 15 N.

Example 4

A person of mass 75 kg enters a lift. He presses the starting button and the lift descends with an acceleration of 1 m s^{-2}. The lift then descends at a steady speed before coming to rest with a deceleration of 1 m s^{-2}.

a) What is the force exerted on the person by the floor when the lift is stationary?
b) What is the force exerted by the floor on the person when the lift is accelerating?
c) Calculate the force exerted by the floor when the lift is decelerating.

a) The weight W of the person is given by

$$W = mg$$
$$= 75 \times 10 = 750$$

This weight of 750 N acts downward
If R is the reaction force exerted by the floor

$$R + W = ma$$

But when the lift is stationary,

$$a = 0$$
$$\Rightarrow R - 750 = 0$$
$$R = 750$$

The force exerted by the floor is 750 N acting upwards.

b) When the lift accelerates downwards

$$a = -1 \text{ m s}^{-2}$$

and $R + W = ma$

$$\Rightarrow R - 750 = 75 \times (-1)$$
$$R - 750 = -75$$
$$\Rightarrow \qquad R = -75 + 750 = 675$$

The force is 675 N acting upwards.

c) The lift is moving downwards but is decelerating. This indicates that there must be an unbalanced force acting upwards. The acceleration is thus directed upwards opposing the downward motion.

In this case $a = +1\,\mathrm{m\,s^{-2}}$

$$R + W = ma$$
$$\Rightarrow R - 750 = 75 \times 1$$
$$R - 750 = 75$$
$$\Rightarrow \qquad R = 75 + 750$$
$$R = 825$$

The force is 825 N and acts upwards.

Example 5

A mass of $0.05\,\mathrm{kg}$ is suspended inside a lift. The lift starts from rest, accelerates upwards at $0.4\,\mathrm{m\,s^{-2}}$, moves upwards at a steady speed of $0.6\,\mathrm{m\,s^{-1}}$ and then decelerates at $0.4\,\mathrm{m\,s^{-2}}$. Find the readings of the spring balance at each stage of the motion.

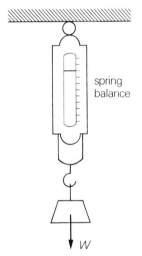

spring
balance

The upward force R is provided by the spring in the balance and the equation relating the forces will be

$$R + W = ma$$

In each stage the magnitude of the weight will be the same and is given by

$$W = mg$$
$$= 0.05 \times 10 = 0.5$$

The weight is $0.5\,\mathrm{N}$

stage 1: the lift at rest

$$R + W = ma \qquad a = 0$$
$$m = 0.05\,\mathrm{kg}$$
$$W = -0.5\,\mathrm{N}$$
$$R - 0.5 = 0.05 \times 0$$
$$R = 0.5$$

The reading on the balance is 0.5 N

stage 2: lift accelerating upwards

$$R + W = ma \qquad a = +\,0.4\,\mathrm{m\,s^{-2}}$$
$$m = 0.05\,\mathrm{kg}$$
$$W = -0.5\,\mathrm{N}$$
$$R - 0.5 = 0.05 \times 0.4$$
$$R - 0.5 = 0.02$$
$$R = 0.02 + 0.5$$
$$R = 0.52$$

The balance reads 0.52 N

stage 3: lift moving at uniform velocity

$$R + W = ma \qquad a = 0$$
$$m = 0.05\,\mathrm{kg}$$
$$W = -0.5\,\mathrm{N}$$
$$R - 0.5 = 0.05 \times 0$$
$$R - 0.5 = 0$$
$$R = 0.5$$

The balance reading is 0.5 N

stage 4: lift decelerating

$$R + W = ma \qquad a = -\,0.4\,\mathrm{m\,s^{-2}}$$
$$m = 0.05\,\mathrm{kg}$$
$$W = -0.5\,\mathrm{N}$$
$$R - 0.5 = 0.05 \times (-0.4)$$
$$R - 0.5 = -0.02$$
$$R = -0.02 + 0.5$$
$$R = 0.48$$

The reading on the balance is 0.48 N.

2.2 Impulse

Suppose that an object, mass m, is acted upon by a net force F for a time t which causes the velocity to change from u to v.

By Newton's Second Law $F = ma$

but the acceleration $a = \dfrac{v - u}{t}$

and substituting for a in the first equation we get

$$F = \frac{m(v - u)}{t}$$
$$= \frac{mv - mu}{t}$$

but mv = final momentum
and mu = initial momentum
so that $(mv - mu)$ is the change in momentum. It follows that

$$\text{net force} = \frac{\text{change in momentum}}{\text{time during which force acts}}$$

and this can be written

$$\text{force} \times \text{time during which it acts} = \text{change of momentum}$$

$$\text{or } Ft = mv - mu$$

The product Ft is called the impulse of the force and the unit of impulse is the newton-second which is the same as the unit of momentum, $\mathrm{kg\,m\,s^{-1}}$.

Forces which act over short time intervals are, in general, not constant. A typical variation of force with time is shown in Figure 2.5. In such a case the impulse is given by the area under the graph, which is indicated by the shading. Thus the change in momentum produced equals the area under the force-time graph.

Often the exact variation of the force is not known and the average force $\bar{F}$ is used so that the equation becomes

$$\bar{F}t = mv - mu$$

where $\bar{F}$ = average force acting during the time interval

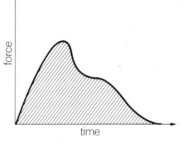

Figure 2.5

Example 6

A billiard cue hits a stationary ball, mass 0.2 kg, which is at rest. The ball moves off with a velocity of $3\,\mathrm{m\,s^{-1}}$ and the time of contact between the ball and the cue is 0.015 s. Calculate the average force exerted by the cue on the ball.

$\bar{F} = \dfrac{mv - mu}{t}$ where $v = 3\,\mathrm{m\,s^{-1}}$
$$u = 0$$
$$m = 0.2\,\mathrm{kg}$$
$$t = 0.015\,\mathrm{s}$$

$$\bar{F} = \frac{0.2 \times 3 - 0.2 \times 0}{0.015}$$

$$= \frac{0.6}{0.015}$$

$$= 40$$

The force exerted by the cue is 40 N.

2.3 Collisions and impulse

When any two objects collide, they exert forces on each other. We can now consider these forces which act when they collide

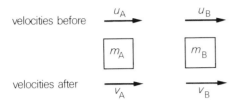

velocities before

velocities after

Figure 2.6

Two objects of mass m_A and m_B collide, and Figure 2.6 shows the various velocities before and after the collision.

During the collision the momentum of the object A changes due to a force $\bar{F}_A$ acting on it for a short time t.

$$\bar{F}_A t = m_A v_A - m_A u_A \quad ...[1]$$

Similarly the momentum of object B changes due to a force $\bar{F}_B$ acting on it for the same time t.

$$\bar{F}_B t = m_B v_B - m_B u_B \quad ...[2]$$

By the law of conservation of momentum

$$m_A u_A + m_B u_B = m_A v_A + m_B v_B$$

$$m_B u_B - m_B v_B = m_A v_A - m_A u_A$$

$$\Rightarrow \quad -(m_B v_B - m_B u_B) = m_A v_A - m_A u_A$$

Hence from equations [1] and [2]

$$-\bar{F}_B t = \bar{F}_A t$$

If the time interval is small, the average force can be regarded as equal to the instantaneous force

$$-F_B = F_A$$

This means that the forces are equal in magnitude but act in opposite directions.

This illustrates Newton's Third Law which can be stated

'to every action force there is an equal and opposite reaction force'

In the collision just described we can write that the force exerted on A by B is equal and opposite to the force exerted on B by A. It is important to realise that action-reaction pairs of forces described by Newton's Third Law do not act on the same object.

2.4 Demonstration of Newton's Third Law

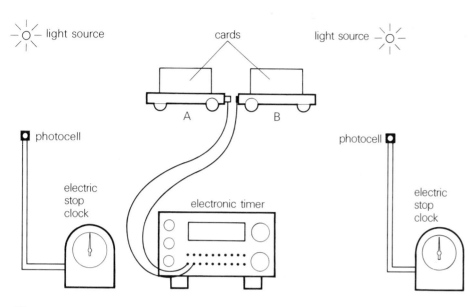

Figure 2.7

A spring-loaded trolley is placed close to but not touching another trolley on a horizontal surface. The metal plunger of trolley A and the metal edge of trolley B are connected to the start terminals of an electronic timer using long thin wire. When the plunger is triggered, it makes contact with the edge of trolley B and the electronic timer records the contact time. Each trolley shoots off and the card attached to it cuts a light beam causing the electric clock to operate. This allows the velocity of each trolley to be calculated.

The following results were obtained.

trolley A

length of card	0.1 m
time on clock	0.32 s
mass m_A	1.20 kg
time of contact	0.05 s

$$\text{speed of A} = \frac{\text{length of card}}{\text{time on clock}}$$

$$= \frac{0.1}{0.32}$$

$$= 0.31 \text{ ms}^{-1}$$

trolley B

length of card	0.1 m
time on clock	0.25 s
mass m_B	0.90 kg
time of contact	0.05 s

$$\text{speed of B} = \frac{\text{length of card}}{\text{time on clock}}$$

$$= \frac{0.1}{0.25}$$

$$= 0.40 \text{ ms}^{-1}$$

We can use these results and the law of conservation of momentum to calculate the average force exerted on each trolley

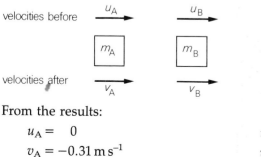

From the results:

$$u_A = 0 \qquad\qquad u_B = 0$$

$$v_A = -0.31 \text{ m s}^{-1} \qquad\qquad v_B = 0.40 \text{ m s}^{-1}$$

To calculate the average force we use $\quad \bar{F} = \dfrac{mv - mu}{t}$

$$\bar{F}_A = \frac{m_A v_A - m_A u_A}{t} \qquad\qquad \bar{F}_B = \frac{m_B v_B - m_B u_B}{t}$$

$$= \frac{1.20 \times (-0.31) - 0}{0.05} \qquad\qquad = \frac{0.90 \times 0.40 - 0}{0.05}$$

$$= -\frac{0.37}{0.05} \qquad\qquad\qquad = \frac{0.36}{0.05}$$

$$= -7.4 \qquad\qquad\qquad\qquad = 7.2$$

$\bar{F}_A$ is 7.4 N $\qquad\qquad\qquad$ $\bar{F}_B$ is 7.2 N

Since we know that
the force of trolley A on trolley B = − (force of trolley B on trolley A)
$$\text{or } \bar{F}_A = -\bar{F}_B$$

we can take the average force to be $\bar{F} = \dfrac{\bar{F}_A + \bar{F}_B}{2}$

$$\Rightarrow \bar{F} = \frac{7.4 + 7.2}{2} = 7.3$$

The average force acting on each trolley will be 7.3 N

2.5 Measuring average force

When a constant force acts on an object
work done on object = force × displacement of the object
$$= Fs$$

Figure 2.8

In many instances the force applied is not constant but varies, as for example, when a stone is fired from a catapult. In these cases an average force $\bar{F}$ must be used.
$$\text{work done} = \bar{F}s$$

This situation can be investigated by using the arrangement shown in Figure 2.9.

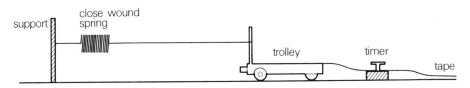

Figure 2.9

A close-wound small spring is attached to a support with the other end fixed to a trolley which is free to move across a horizontal surface. A length of string is used to join the spring to the trolley so that the trolley is free to run for some distance before hitting the spring. The trolley is pulled away from the support until the spring is stretched by 0.095 m. The trolley is released and accelerates. The distances travelled by the trolley are recorded on a ticker tape. The frequency of the timer is 50 Hz and the mass of the trolley is 0.84 kg.

The average force acting on the trolley can be found in three ways:

a) by direct measurement
b) by considering energy
c) by considering impulse

Each method will now be considered in detail.

By direct measurement
A spring balance reading up to 5 newtons is used to measure the force acting on the spring when it is extended, Figure 2.10.

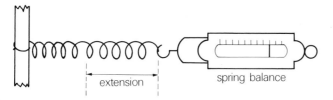

Figure 2.10

The results obtained were as follows.

force (N)	0.5	1.0	1.5	2.0	2.5	3.0
extension (m)	0.019	0.040	0.058	0.076	0.096	0.113

This gives the graph shown in Figure 2.11.

The force and extension for the spring just before the trolley is released is shown with dotted lines on the graph.

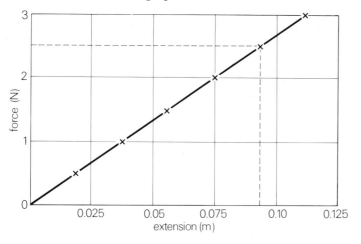

Figure 2.11 Graph of force against extension for the spring

The graph illustrates that the extension is directly proportional to the applied force. In this case the average force $\bar{F}$ is given by

$$\bar{F} = \frac{\text{final force} + \text{initial force}}{2}$$

The initial extension of the spring during the experiment was 0.095 m. From the graph the final force is thus just under 2.5 N.
The initial force (unstretched spring) is zero.

The average force is $\bar{F} = \dfrac{2.5 + 0}{2} = 1.3$

The average force estimated by direct measurement is 1.3 N

By considering energy

The tape attached to the trolley is shown full size (Figure 2.12)

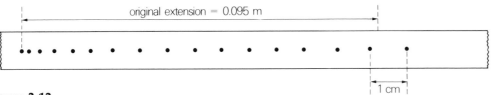

Figure 2.12

The potential energy stored in the stretched spring is transferred to the trolley which gains kinetic energy. The initial extension of the spring is important because after the trolley has travelled this distance the spring will no longer exert a force on the trolley. The velocity at this point will thus be a maximum.

final kinetic energy of the trolley = average force × displacement

$$\tfrac{1}{2}mv^2 = \bar{F}s$$

From the tape it can be seen that the spring will cease to exert a force during the sixteenth space. The maximum velocity will therefore occur during this time interval.

The velocity is given by $v = \dfrac{\text{length of space}}{\text{time interval}} = \dfrac{0.01}{0.02} = 0.5$

The maximum velocity of the trolley is $0.5\,\text{ms}^{-1}$

The kinetic energy of the trolley $E_k = \tfrac{1}{2} \times 0.84 \times 0.5^2 = 0.11$

The kinetic energy of the trolley is $0.11\,\text{J}$
But the kinetic energy gained by the trolley $= \bar{F}s$ and since the original extension of the spring is 0.095

$$\bar{F} \times 0.095 = 0.11$$

$$\bar{F} = \frac{0.11}{0.095} = 1.2$$

The average force calculated by considering energy is 1.2 N

By considering impulse

impulse = change of momentum

$$\bar{F}t = mv - mu$$

where the time during which force acts $= t$ initial velocity of trolley $= u$
                       final velocity of trolley $= v$ mass of trolley $= m$

From the tape it can be seen that the spring ceases to exert a force on the trolley during the sixteenth time interval.

$$t = 16 \times 0.02 = 0.32$$

The spring exerts a force for a time of $0.32\,\text{s}$
The velocity v (calculated in the previous section) $= 0.5\,\text{m s}^{-1}$ and the initial velocity u is zero.

$$\text{impulse} = \text{change of momentum}$$
$$\Rightarrow \bar{F} \times 0.32 = 0.84 \times 0.5 - 0.84 \times 0$$
$$\bar{F} \times 0.32 = 0.42 \qquad \Rightarrow \bar{F} = \frac{0.42}{0.32} = 1.3$$

The average force calculated by considering impulse is 1.3 N

It can be seen that all methods give substantially the same result.

Example 7

Two trolleys A (mass 1 kg) and B (mass 3 kg) are attached with an elastic cord and pulled apart by a distance of 1 m on a horizontal surface.

They are released simultaneously and collide at point P. Find the displacements s_A and s_B.

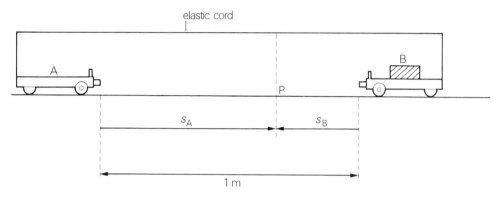

Because of Newton's Third Law the average force $\bar{F}_A$ exerted on trolley A by the elastic cord has the same magnitude as the average force $\bar{F}_B$ exerted on trolley B by the elastic cord, but it will act in the opposite direction.

$$\bar{F}_A = -\bar{F}_B$$

The work done on trolley A by the cord $= \bar{F}_A s_A$

This will provide the trolley with kinetic energy given by $\frac{1}{2}mv_A^2$ where v_A = velocity of A just before impact

$$\bar{F}_A s_A = \frac{1}{2}mv_A^2$$

In the same way $\bar{F}_B s_B = \frac{1}{2}mv_B^2$

but $m_A = 1$ kg and $m_B = 3$ kg $\Rightarrow$ $\bar{F}_A s_A = \frac{1}{2}v_A^2$ ... [1]

$\Rightarrow$ $\bar{F}_B s_B = \frac{3}{2}v_B^2$ ... [2]

The initial momentum is zero and since there is no external force acting on the whole system, the momentum just before impact is also zero.

$$m_A v_A + m_B v_B = 0$$

$$m_A v_A = -m_B v_B \quad \Rightarrow \quad v_A = -3v_B$$

If we divide the energy equations [1] and [2] and substitute $-3v_B$ for v_A we have

$$\frac{\bar{F}_B s_B}{\bar{F}_A s_A} = \frac{\frac{3}{2}v_B^2}{\frac{1}{2}v_A^2}$$

$$\frac{\bar{F}_B s_B}{\bar{F}_A s_A} = \frac{3v_B^2}{v_A^2} \quad \Rightarrow \quad \frac{-\bar{F}_A s_B}{\bar{F}_A s_A} = \frac{3v_B^2}{(-3v_B)^2}$$

$$\frac{-s_B}{s_A} = \frac{3v_B^2}{9v_B^2} \quad \Rightarrow \quad \frac{-s_B}{s_A} = \frac{1}{3}$$

$$s_A = -3s_B$$

The total distance between the trolleys at the start is 1 m.

The magnitude of s_A must be 0.75 m and of s_B must be 0.25 m.

If we wish to indicate the vector direction of the displacements, we can write

$$s_A = +0.75 \text{ m and } s_B = -0.25 \text{ m}$$

2.6 Force as a vector

The quantities displacement, velocity and force are vector quantities and are therefore added as vectors. The combined effect of two or more vectors is called the resultant of the vectors and is their vector sum. If two forces act on an object as shown in Figure 2.13 the resultant can be found using a scale diagram in which the length of the line represents the magnitude of the force and the angle at which the line is drawn represents the direction. The vectors are joined head to tail as indicated, Figure 2.14.

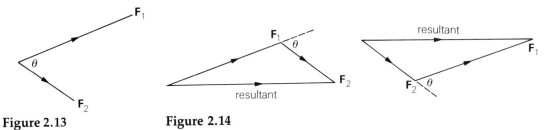

Figure 2.13   **Figure 2.14**

The same value for the magnitude and direction of the resultant is obtained irrespective of which vector is drawn first.

Example 8

Two forces of 15 N and 6 N act on an object at an angle of 80° between the directions of the forces. Find the resultant force. For the scale diagram choose 1 cm to represent a force of 2 N.

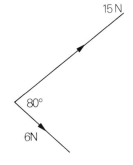

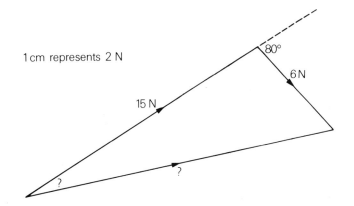

The length of the resultant vector is 8.6 cm so the resultant force is 8.6×2 which is 17.2 N. The angle between the resultant force and the 15 N is measured to be 20°.

Resolution of vectors

Two forces can be combined to give a single force called the resultant which as we have seen can replace these two forces. The reverse process is also possible, a single force being replaced by two forces called components. This splitting up into two separate forces is known as **resolution** and the components are at right angles to each other.

This is shown in Figure 2.15 and as indicated in the vector triangle in Figure 2.16, F_x and F_y combined form the single force **F**.

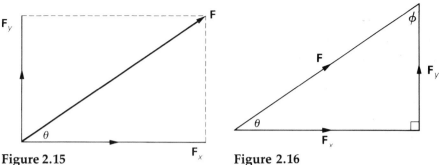

Figure 2.15

Figure 2.16

From the triangle, $\dfrac{F_y}{F} = \cos \phi$ and $\dfrac{F_x}{F} = \cos \theta$

$\Rightarrow F_y = F \cos \phi \qquad \Rightarrow \qquad F_x = F \cos \theta$

Example 9

A sledge mass 8 kg is pulled across a horizontal surface by a force of 10.4 N acting at an angle of 30° to the horizontal. If friction can be neglected, calculate the acceleration of the sledge. Only the horizontal component of the force will accelerate the sledge across the surface.

horizontal component $= 10.4 \cos 30° = 9$

the acceleration is given by $\mathbf{F} = m \times \mathbf{a}$

$$\Rightarrow \mathbf{a} = \frac{\mathbf{F}}{m}$$

$$\mathbf{a} = \frac{9}{8} = 1.12$$

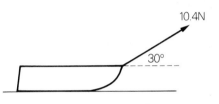

The acceleration of the sledge is 1.12 m s⁻²

Example 10

An object of mass 25 kg rests on an inclined plane which is at an angle of 40° to the horizontal. If the force of friction is 40 N, calculate the acceleration of the object down the plane.
  The weight W of the object acts vertically downwards and is given by

$$W = m \times g$$
$$= 25 \times 10 = 250$$

From the diagram we can see that the weight makes an angle of 50° to the plane.
  Since we require the acceleration down the plane, the force which produces this will be the component of the weight acting down the plane.

component down plane $= W \cos 50°$

$$= 250 \times 0.64 = 160.1$$

But the friction force also acts along the plane but upwards

therefore net force $\quad = 160.1 - 40 = 120.1$

the acceleration is found using $F = m \times a$

$$a = \frac{F}{m}$$

$$= \frac{120.1}{25} = 4.8$$

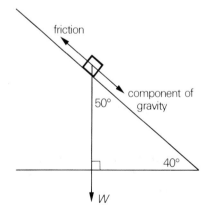

The acceleration of the object down the slope is 4.8 m s⁻²

Summary

Newton's Second Law is expressed by the equation

$$F = m a$$

The unbalanced force acting on an object is equal to the rate of change of momentum

$$F = \frac{mv - mu}{t}$$

The impulse of a force is equal to the product of the force and the time over which it acts.

$$\text{Impulse} = F t$$

When a force is not constant, the impulse is equal to the area under the force-time graph

A vector can be resolved into two components at right angles to each other.

Problems

1 Describe an experiment which could be conducted to estimate the unbalanced force acting on a trolley which is accelerated uniformly along a horizontal surface by a stretched elastic cord. How could you check the answer you obtain for the force?

2 A train, mass 250 tonnes, experiences a total frictional resistance of 10 kN as it accelerates at 2 m s^{-2} along a horizontal track.
   a) Calculate the total driving force required to produce this acceleration.
   b) Assuming that the frictional resistance remains the same, calculate the driving force required to maintain the same acceleration up an incline which makes an angle of 8° to the horizontal.

3 The string of a pendulum is held at an angle of 45° as shown. The mass of the bob is 0.1 kg and the length of the pendulum is 1 m.

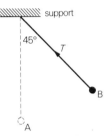

   a) How much potential energy has been gained by the bob in being lifted from A to B?
   b) What is the tension T in the string when it is let go?
   c) Determine the initial acceleration of the bob when it is let go.
   d) How does the tension in the string at A compare with the tension at B?

4 An object, mass 5 kg, falls freely towards the surface of a planet with an acceleration of 5 m s^{-2}.
   a) What is the force of gravity, acting on the object, due to the planet?
   b) At what height will the object have gained 150 joules of potential energy if it were raised from the surface of the planet?

5 An elastic cord is stretched between two supports as shown.

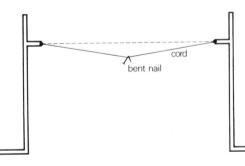

A bent nail is suspended on the cord; the cord is released and the nail is catapulted vertically upwards.
Explain how you would estimate the potential energy stored in the cord.

6 a) Which physical quantity is defined as the change in momentum per unit time?
b) In which unit is this quantity measured?

7 a) A golf club exerts an average force of 2.8 kN on a ball of mass 0.05 kg. If the contact time is 4×10^{-4} s, determine the velocity of the ball as it leaves the club.
b) A vehicle, mass 0.1 kg, moves along a horizontal linear air track at a constant speed of 0.2 m s^{-1}. If it is in contact with an elastic cord for 0.3 s and rebounds with a speed of 0.16 m s^{-1}, calculate the average force acting on the vehicle.

8 Which physical quantities are represented by the shaded areas shown in the graphs?

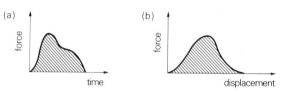

9 Two trolleys are exploded apart on a horizontal surface. If trolley A moves off at 1.5 m s^{-1}, calculate the velocity of B. Calculate the kinetic energies of the trolleys just as they move off.

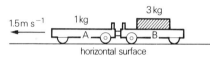

10 A ball mass 0.1 kg is dropped from a height of 3.2 m.
a) How long will it take to hit the ground?
b) Construct a graph showing how the kinetic energy varies with time until the ball hits the ground.

11 A spring is attached to the lid of a cocoa tin and a weight of 1 N is hung at the end. A pointer is fixed to the spring and sticks out from a slit which is cut in the side of the tin. Explain what you would expect to happen to the pointer in the following situations.
a) the tin is allowed to fall down vertically
b) the tin is thrown vertically upwards
c) the tin is placed on a trolley which is then pushed along a horizontal surface

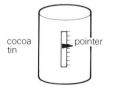

12 The force acting on an object, mass 1.5 kg, varies with time as shown.

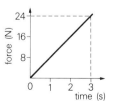

a) Determine the impulse of the force over this 3 second period.
b) What is the change in momentum of the object after 3 s?
c) Calculate the kinetic energy of the object at the end of 3 s if the initial velocity of the object was zero.

13 A rocket-powered sledge has a mass of 210 kg unloaded. During a test run an astronaut, mass 85 kg with equipment, sits on the sledge and is accelerated along a frictionless horizontal track. The speed is measured and automatically recorded, the results of which are given in the graph.

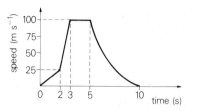

a) What distance was covered in the first 5 seconds?
b) Calculate the thrust provided during the first 2 seconds.
c) Was the decelerating force acting on the sledge constant?

14 The unbalanced thrust on an object, mass 100 kg, varies as shown.

a) If the initial speed of the object is zero, draw a speed-time graph for the object for the 20 second period.
b) Calculate the distance travelled by the object during this time.

15 A boxer exerts an average force of 200 N on a punchbag for a time of 0.15 s. The mass of the bag is 40 kg.
a) What is the velocity of the bag after it leaves the glove?
b) How high would the bag rise vertically above the starting position?
c) Describe a method by which the average force exerted by a boxing glove could be estimated.

16 When a catapult is used to fire a stone, it exerts an average force of 12 N. The stone has a mass of 0.043 kg and acquires a speed of 7 m s^{-1}. Estimate the time of contact between the catapult and the stone. How could the average force exerted by the elastic of the catapult be estimated?

17 An electric train has a mass of 2.5×10^5 kg. It makes a journey between two stations on a horizontal track. The speed changes with time as indicated on the graph. The effective resistance opposing the motion of the train is 20 kN.

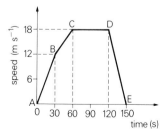

a) Calculate the acceleration during each section of the motion.
b) Determine the total force applied during section BC.
c) What is the driving power of the train during period CD?

2 Forces

18 Two blocks slide towards each other, collide and stop.

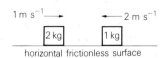

a) Is momentum conserved?
b) Calculate the loss of kinetic energy.
c) Where has this kinetic energy gone?

19 An object, mass 0.8 kg, slides from rest down an incline. It slides 2 m down the incline, falling a vertical height of 1.5 m and attaining a speed of 2 m s^{-1}.

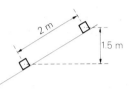

a) Calculate the loss of potential energy.
b) What is the kinetic energy of the object?
c) Determine the force of friction acting on the object.

20 The motion of a 5 kg radio-controlled car is represented on the graph.

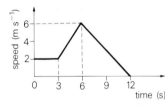

a) What was the initial speed of the car?
b) Draw a graph showing the unbalanced force acting on the car during the 12 seconds shown.
c) Calculate the distance travelled by the car.

21 A vehicle on a linear air track has a mass of 80 g. It is catapulted by an elastic cord which is pulled back a distance of 0.09 m. When released the vehicle moves off at 0.4 m s^{-1} after being in contact with the cord for 0.045 s.
a) Calculate the momentum of the trolley.
b) What is the average force exerted by the cord?
c) Estimate the original potential energy stored in the stretched cord.

22 A student of mass 70 kg investigates the motion of a lift. He stands on a weighing machine in the lift on its downward journey in a high building. For 2 seconds immediately after the lift starts, the weighing machine reads 560 N; then for a further 6 seconds it reads 700 N, and for the final 2 seconds it reads 840 N.
a) Describe the motion of the lift during its journey.
b) Calculate the magnitude and direction of the resultant force acting on the student during each stage of the journey.
c) Draw a graph of acceleration against time for the journey. The axes of the graph should clearly indicate the values of the acceleration at different times.

SCEEB

23 In an experiment to check the calibration of a spring balance a trolley of mass 4.0 kg is pulled down a friction-compensated track. Positions of the marker straw attached to the trolley are obtained from a stroboscopic photograph.

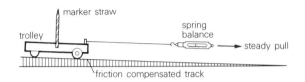

A section of the resulting stroboscopic photograph, enlarged to true size, is shown. The flash rate of the stroboscope was 10 flashes per second and the spring balance registered a steady reading of 0.90 N.

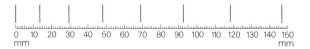

a) Determine the resultant force applied to the trolley.
b) Suggest **two** reasons which might account for the difference between the calculated resultant force and the spring balance reading. Explain how each reason accounts for the observed difference.

SCEEB

24 A vehicle is travelling along a horizontal linear air track. A card on the vehicle passes through a light beam before the vehicle is stopped by a stretched elastic band. While the vehicle is being stopped, the centre of the elastic moves from O to P.

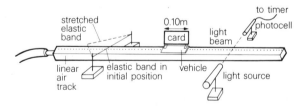

Results obtained from one run of the vehicle are:

Mass of vehicle and card = 1.1 kg

Length of card = 0.10 m

Time through light beam = 0.25 s

Stopping distance OP = 0.050 m

a) Calculate
  i) the average force exerted by the elastic band on the vehicle while it is being stopped;
  ii) the time taken by the elastic band to stop the vehicle.
b) Describe how you would measure the maximum stretch OP of the elastic band experimentally.

SCEEB

25 In driving a pile into the ground, a hammer of mass 500 kg falls freely from rest through a height of 5.0 m on to a pile of mass 1500 kg. The pile and hammer then move together as the pile is driven 0.12 m into the ground.

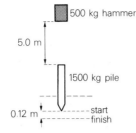

500 kg hammer

5.0 m

1500 kg pile

0.12 m — start — finish

a) Determine the speed of the hammer just before it hits the pile.
b) i) Using the Principle of Conservation of Momentum, calculate the common speed of pile and hammer immediately after the collision.
 ii) State one assumption which you must make to justify your application of momentum conservation in part (*b*) (i).
c) From the moment just after collision until the system comes to rest, what is the change in
 i) the total kinetic energy of pile and hammer:
 ii) the total potential energy of pile and hammer?
d) By considering these energy changes, or otherwise, calculate a value for the average resistive force which the ground offers to the motion of the pile during its movement into the ground.

SCEEB

26 A train consists of an engine and a line of three wagons. Each wagon has a mass of 20 000 kg. The resistive forces due to friction acting on the wagons can be assumed constant at all speeds and equal to 1000 N on each wagon.

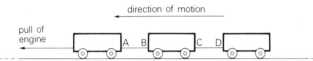

direction of motion

pull of engine

A B   C D

a) When the train is moving along a straight, horizontal stretch of track, the engine exerts a constant pull of 45 000 N on the front wagon.
Calculate **i)** the acceleration of the train;
 ii) the tension in the coupling chain, AB.
b) With the same engine pull continuing to act, the train moves on to a straight, downhill section, the slope of which just compensates for the resistive forces acting on the wagons. During this part of the run state whether the tension in the coupling chain CD is less than, equal to, or greater than the tension in coupling chain AB. **Explain** your choice.
c) Finally the train moves on to another straight, horizontal track and the engine pull is reduced until the train is running at a constant speed. Again **explain** how the tension in coupling chain CD will compare with the tension in coupling chain AB.

SCEEB

27 In a laboratory experiment a vehicle on a track is set in motion by a catapult incorporating a number of identical elastic cords. The diagram shows the arrangement of the catapult system.

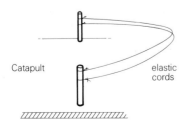

Catapult

elastic cords

a) What practical steps would you take to ensure that friction is made negligible and how would you check that this has been done?
b) The speed achieved by the vehicle when catapulted by different numbers of elastic cords is recorded in the table below.

Number of cords N	Speed in m/s v
1	0.41
2	0.57
3	0.69
4	0.80

 i) From theoretical considerations, what relationship would you expect between N and v?
 ii) Show whether the results above verify this relationship.
c) State any **two** practical factors likely to lead to discrepancies in the results.

SCEEB

28 In an experiment, a block of mass 1.00 kg is released from rest from a point A 2.00 m up a slope as shown in the diagram. The block slides down to the point B at the bottom of the slope where its speed is measured. This is repeated several times, the block being released from the same point A each time but with the slope adjusted so that the initial height h is different on each occasion.

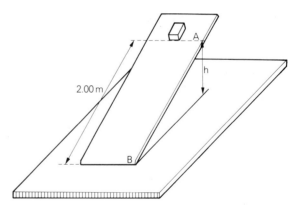

A

h

2.00 m

B

The speeds of the block for four different initial heights are shown in the table.

Height h (m)	0.60	1.00	1.40	1.80
Speed at B (m s^{-1})	2.50	3.85	4.90	5.80
Potential energy at A (J)				
Kinetic energy at B (J)				

a) i) Copy the table and complete it to show the potential energy of the block at A and the kinetic energy of the block at B.
 ii) Account for the difference between the potential energy at A and the corresponding kinetic energy at B.
b) Calculate the average frictional force acting on the block during the experiment when $h = 1.00$ m.
c) i) Plot a graph of the kinetic energy at B against the initial height h.
 ii) Use this graph to find a value for the initial height h which makes the slope friction-compensated for this block.

SCEEB

3 d.c. electricity

3.1 Revision

Current

More than 160 years ago, 'electric current' was just beginning to be investigated scientifically. In 1819 a Danish scientist called Hans Christian Oersted showed that a magnetic field was produced around a wire through which a current was being passed. This field can be detected by the movement of a magnetic compass needle placed nearby. The direction of the current governs the direction of the magnetic field and is given by the **left-hand rule**, Figure 3.1.

Note that in this text we shall refer to the **direction** of current as that of the **flow of electrons** (many texts refer to the direction of conventional current which is in the reverse direction).

A wire carrying a current at right angles to a magnetic field experiences a force which tends to make it move in a direction given by the **right-hand motor rule**, Figure 3.2. The behaviour was used by André Ampère, the French scientist to investigate the factors governing the strength of this force.

Ampère realised that two parallel wires, each carrying a current, would produce two magnetic fields. These fields would exert a force on the wires if the wires were near to each other, Figure 3.3.

He made measurements from which he concluded that the force F between the two wires of length l depended on the currents I_1 and I_2 in each wire, on the distance d between the two wires and on the length l of the wires. The force was attractive if the currents were in the same direction and repulsive if they were in the opposite direction.

The results of his experiment are summarized as follows:

force varies directly as the current in the first wire $F \propto I_1$

force varies directly as the current in the second wire $F \propto I_2$

force varies directly as the length of wire considered $F \propto l$

force varies inversely as the distance between the wires $F \propto \dfrac{1}{d}$

Combining them, we obtain

$$F \propto \frac{I_1 I_2 l}{d}$$

$$\Rightarrow \frac{F}{l} \propto \frac{I_1 I_2}{d}$$

$$\Rightarrow \frac{F}{l} = \frac{k I_1 I_2}{d} \text{ where } k \text{ is a proportionality constant.}$$

The ampere (A) is the unit of current and is defined as follows:

> The ampere is the constant current which, when flowing in two infinitely long straight parallel conductors of negligible cross section placed one metre apart in a vacuum, produces between them a force of $2 \times 10^{-7}\,\text{N}$ for each metre length of wire.

From this definition, for values of $I_1 = I_2 = 1\,\text{A}$, $l = 1\,\text{m}$ and $d = 1\,\text{m}$, we find that the force is $2 \times 10^{-7}\,\text{N}$. Substituting these values in the equation, we see that $k = 2 \times 10^{-7}$.

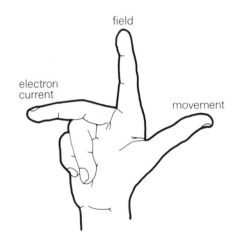

Figure 3.1 Left-hand rule

Figure 3.2 Right-hand motor rule

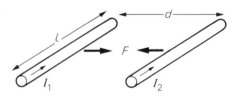

Figure 3.3 Forces due to currents

Charge

A current is produced when an electric charge moves round a circuit. The rate of flow of charge is defined as the current.

$$\frac{Q}{t} = I$$

where Q is a quantity of charge flowing for time t, and I is the current.

$$\Rightarrow \quad Q = It$$

Thus, for a current of 1 ampere, the quantity of charge that flows past a given point in 1 second is defined as 1 coulomb (C), or in terms of units

$$1\,C = 1\,A \times 1\,s$$

1 coulomb = 1 ampere-second

Example 1

A current of 5 mA flows in a circuit for 2 s. How much charge has passed a given point in the circuit?

$$I = 5\,mA = 5 \times 10^{-3}\,A$$
$$t = 2\,s$$

Using $\quad Q = I \times t$

$$\Rightarrow \quad Q = 5 \times 10^{-3} \times 2 = 10 \times 10^{-3}$$
$$= 10 \text{ millicoulombs}$$

A charge of 10 mC has passed.

Electrical conduction

A material through which electric charge can flow is called a **conductor**. A material through which electric charge will *not* flow is called an **insulator**. Conduction may take place through gases, liquids or solids, but we shall only consider solids.

Most solid conductors are metals. It is useful to build up a model of the process of conduction. Figure 3.4 depicts a piece of metal with one 'free electron' per atom. The black circles represent the nuclei and fixed electrons of the metal atoms. Atoms have nuclei surrounded by electrons, most of which are bound tightly to the nucleus. However in metal atoms, some of the outer electrons are free to move. These are shown in the diagram and are called 'free electrons'. The American scientist Edwin Hall showed experimentally that in most metals one or two 'free electrons' per atom are available for conduction of electricity.

These electrons have relatively high speeds (about $10^6\,m\,s^{-1}$) in random directions. When a conductor is placed in a circuit containing a battery, this general movement becomes directed towards the positive terminal, Figure 3.6. There is a drift of all the free electrons at a speed of about $10^{-4}\,m\,s^{-1}$ as soon as the circuit switch S is closed. The electron current is from negative to positive terminal (conventional current is in the reverse direction).

Potential difference and resistance

The energy required to drive the electron current round the circuit is provided by a chemical reaction in the battery or by the mains power supply. The **potential difference** across any component X in a circuit can be measured by placing a voltmeter in parallel with it, Figure 3.7. The component X is said to have a **resistance** if a potential difference (p.d.) is needed in order to drive a current through it.

Figure 3.4 Metal atoms

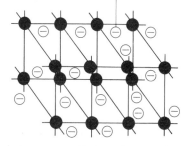

Figure 3.5 Edwin Hall

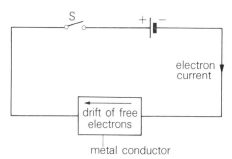

Figure 3.6 Conduction in a metal

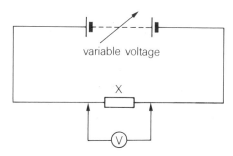

Figure 3.7 Measuring potential difference

The graph in Figure 3.8 shows the relationship usually obtained between the current I through a resistor (component X) and the potential difference V across it. This shows that the current varies directly as the potential difference.

$$I \propto V \text{ or } V \propto I$$

$\Rightarrow \quad V = \text{constant} \times I$

$\Rightarrow \quad \dfrac{V}{I} = \text{constant} = R$

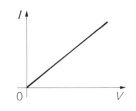

Figure 3.8 Current-voltage graph for a resistor

where the constant R is called the resistance. The ratio V/I is the resistance of the component.

The resistance can be thought of as a measure of the potential difference needed to drive a current of 1 ampere through the component. If 1 volt is required to drive a current of 1 ampere through a component, it is said to have a resistance of 1 ohm.

$$R = \dfrac{V}{I}$$

If $V = 1$ volt, $I = 1$ ampere, then $R = 1$ ohm

$1 \text{ ohm} = 1$ volt per ampere

$1\Omega = 1 \text{ V A}^{-1}$

Example 2

What potential difference is required to drive a current of 1 A through a resistor a) of value 2Ω b) of value 120Ω?

A 1Ω resistor requires 1 V to drive a current of 1 A
A 2Ω resistor requires ? V to drive a current of 1 A
– twice the resistance: twice as difficult, so twice the voltage needed, i.e. 2 V.

So, the 120Ω resistor requires 120 V to drive a current of 1 A.

The electrical energy in a circuit, supplied by the source, is converted to other forms of energy in the components which make up the circuit. The amount of electrical energy converted into other forms of energy when the coulomb of charge passes from one point in the circuit to another is called the **potential difference** between the two points.

If 1 coulomb of charge passes between two points A and B in a circuit and releases 1 joule of energy, then we say that there is a potential difference of 1 volt between A and B, Figure 3.9.

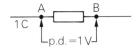

Figure 3.9 Source connected to a resistor

If the p.d. is 2 volts, then 2 joules of energy is released per coulomb of charge which passes between A and B.

If the p.d. is V volts then V joules of energy is released per coulomb of charge which passes between A and B.

If Q coulombs of charge pass between A and B when the p.d. is V volts then the energy released, W, is given by $W = QV$.

If charge passes as an electric current I, where $I = Q/t$, then we can express the energy released by the equation

$$W = QV = I t V \text{ joules}$$

Example 3

Calculate the energy converted in a 4 ohm resistor when connected to a 12 volt source for 2 minutes.

$$R = \frac{V}{I} \implies 4 = \frac{12}{I} \implies I = 3 \text{ amperes}$$

$$W = I\,t\,V \qquad t = 2 \times 60 \text{ seconds}$$

$$\implies W = 3 \times 2 \times 60 \times 12 = 4320$$

4320 joules of energy are converted in the 4 ohm resistor

3.2   Electromotive force

A source of electrical energy, such as a battery, provides an electromotive force (e.m.f.) to drive an electron current through the circuit. The e.m.f. of the source governs the electrical energy supplied to each unit charge that passes through the source, i.e. the number of joules of energy per coulomb of charge. Alternatively, the e.m.f. of the source is the work done on unit charge when it passes through the source.

The units of e.m.f. are the same as those of p.d.

$$1 \text{ J C}^{-1} \equiv 1 \text{ V}$$

The e.m.f. of a source is the potential difference across the terminals when no current is flowing, i.e. on open circuit. If a current is flowing, the terminal potential difference (t.p.d.) is lower than the e.m.f. as explained below.

Internal resistance

Consider the circuit in Figure 3.10. We might expect the potential difference V across the resistor R to be the same as the e.m.f. of the source. In reality a set of results for various values of R is obtained like those in the table below.

resistor value $R\ \Omega$	100	10	5	2	1
voltmeter reading V volts	2.0	1.7	1.4	1.0	0.7

Table 1

We find that the value of the external resistance R affects the terminal potential difference. This is because the cell used as a source has an **internal resistance** which resists the current.

Measurement of internal resistance

The circuit in Figure 3.11 is used for determining the value of the internal resistance r of a cell of e.m.f. 1.8 V by measuring the current and the t.p.d. for various values of external resistance R.

Table 2 below shows a typical set of results.

current I amperes	0.0	0.1	0.2	0.3	0.4	0.5	0.6	0.7	0.8
t.p.d. V volts	1.8	1.6	1.4	1.2	1.0	0.8	0.6	0.4	0.2

Table 2

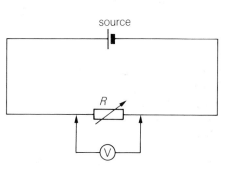

Figure 3.10 Measuring internal resistance

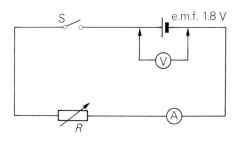

Figure 3.11

These results are plotted on a graph of t.p.d. against current, Figure 3.12. Notice that when switch S is open and no current flows, the t.p.d. has a maximum value which is equal to the e.m.f. of the cell: the cell is on 'open circuit'.

The graph is a straight line of the form

$$y = mx + c$$

where m is the gradient of the line and c is the intercept on the vertical axis.

$$\Rightarrow V = m\,I + c$$

$$\Rightarrow V = -2.0\,I + 1.8$$

because the gradient of the line is -2.0 and the intercept on the vertical axis is $+1.8\,V$ (the e.m.f. E).

$$\Rightarrow V = E - Ir \text{ where } r = -m, \text{ the gradient.}$$

The terminal potential difference V is less than the e.m.f. by some quantity Ir where r is the internal resistance of the cell. The value of Ir is sometimes called the 'lost volts'.

Because $m = -2.0$, the internal resistance r of the cell is $2.0\,\Omega$.

Energy is lost in driving the current through the chemicals which make up the cell, and this causes the cell to have an internal resistance r.

Note also the value $0.9\,A$ of the intercept on the horizontal axis. This shows that the maximum current which the cell is capable of delivering (i.e. when the cell is on short-circuit) is $0.9\,A$.

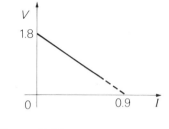

Figure 3.12

Example 4

A circuit contains a battery of e.m.f. $4.0\,V$ and internal resistance $1.5\,\Omega$. If the external resistor has a value of $6.5\,\Omega$, find the value of

a) the current in the circuit
b) the t.p.d.
c) the short-circuit current.

In the circuit on the right, the battery is shown as an e.m.f. of $4.0\,V$ and an internal resistance r of $1.5\,\Omega$ in series.

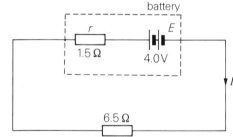

a) Total circuit resistance $= 6.5 + 1.5 = 8.0\,\Omega$
e.m.f. of battery $= 4.0\,V$

$$I = \frac{V}{R}$$

$$\Rightarrow I = \frac{4.0}{8.0} = 0.5$$

The current in the circuit is 0.5 A

b) The t.p.d. is equal to the p.d. across the external resistor.

$$V = I\,R$$

$$\Rightarrow V = 0.5 \times 6.5 = 3.25$$

The t.p.d. is 3.25 V

c) The short-circuit current is the current when the voltage V is zero.

$$V = E - Ir$$

$$\Rightarrow \quad 0 = 4.0 - I \times 1.5$$

$$\Rightarrow 1.5 \times I = 4.0$$

$$\Rightarrow \quad I = \frac{4.0}{1.5} = 2.7$$

The short-circuit current is 2.7 A.

Example 5

Calculate the internal resistance of each of the following sources of e.m.f. when inserted in the circuit shown on the right.

a) a U2 dry cell, e.m.f. = 1.5 V, t.p.d. = 1.2 V
b) a car battery, e.m.f. = 12.0 V, t.p.d. = 11.7 V
c) a calculator battery, e.m.f. = 9.0 V, t.p.d. = 3.6 V

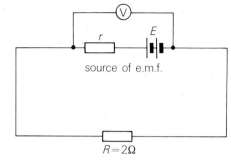

$$V = E - Ir \quad \text{where} \quad r \text{ is internal resistance of source.}$$

$$\text{but} \qquad I = \frac{E}{R + r} \qquad R \text{ is external resistance}$$

$$\Rightarrow \qquad V = E - \left(\frac{E}{R + r}\right) r$$

$$\Rightarrow V(R + r) = E(R + r) - Er$$

$$\Rightarrow VR + Vr = ER$$

$$\Rightarrow \qquad r = \frac{R(E - V)}{V}$$

a) For the U2 dry cell $E = 1.5$ V, $V = 1.2$ V, $R = 2\,\Omega$

$$\therefore \qquad r = \frac{2(1.5 - 1.2)}{1.2} = 0.5$$

Internal resistance of U2 cell = 0.5 Ω

b) For the car battery $E = 12.0$ V, $V = 11.7$ V, $R = 2\,\Omega$

$$\therefore \qquad r = \frac{2(12.0 - 11.7)}{11.7} = 0.05$$

Internal resistance of car battery = 0.05 Ω

c) For the calculator battery $E = 9.0$ V, $V = 3.6$ V, $R = 2\,\Omega$

$$\therefore \qquad r = \frac{2(9.0 - 3.6)}{3.6} = 3.0$$

Internal resistance of calculator battery = 3.0 Ω.

3.3   Power in a d.c. circuit

The rate at which a component converts electrical energy into another form is known as its power, P. The unit of power is the watt. The energy converted in a resistor was given in section 3.2 as

$$W = I t V$$

Thus the rate of energy conversion $= \dfrac{W}{t} = \dfrac{I t V}{t}$

$$\Rightarrow \text{power} = I V \text{ watts}$$

The internal resistance of a source results in non-useful energy conversion within the source so that the power available to the external circuit is reduced. The circuit in Figure 3.13 is used to investigate power transfer in an electric circuit.

By varying the value of the external resistor R and measuring the correspond-ing current and p.d. values, it is possible to determine the power delivered to

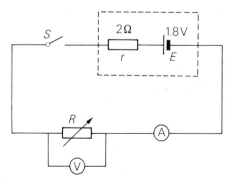

Figure 3.13 Power transfer

3   d.c. electricity

the external circuit. Table 3 shows a typical set of results. The external power has been calculated from the product $I \times V$.

external resistance, R (Ω)	7.0	4.0	2.5	1.6	1.0	0.6
current, I (A)	0.2	0.3	0.4	0.5	0.6	0.7
p.d. across R, V (V)	1.4	1.2	1.0	0.8	0.6	0.4
power, P, (W)	0.28	0.36	0.4	0.4	0.36	0.28

Table 3

A graph of power P supplied to the external circuit against value of external resistor R is show in Figure 3.14.

This graph indicates that the power supplied to the external circuit has a **maximum** value when the external resistance is 2 ohms. This corresponds to the internal resistance of this source.

Maximum transfer of power occurs when the internal resistance of the source is equal to the external resistance of the circuit being supplied.

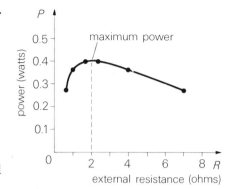

Figure 3.14

Example 6

A cell of internal resistance 0.5 ohms and e.m.f. 1.5 volts is used to supply power to a torch bulb. What is the maximum power of bulb which should be used?

For maximum power $R = r = 0.5\,\Omega$

$$I = \frac{E}{R + r}$$

$\Rightarrow$
$$I = \frac{1.5}{0.5 + 0.5} = 1.5\,\text{amperes}$$

p.d. across lamp $V = E - Ir$

$\Rightarrow$
$$V = 1.5 - 1.5 \times 0.5$$
$$= 0.75$$

p.d. across lamp $= 0.75$ volts

Maximum power $P = V \times I$
$$= 0.75 \times 1.5 = 1.125$$

The maximum power of bulb is 1.125 watts.

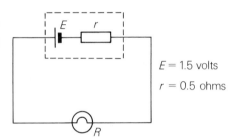

$E = 1.5$ volts
$r = 0.5$ ohms

Summary

A straight current-carrying wire at right angles to a magnetic field experiences a force at right angles to both the wire and the field. The direction of the movement that this force tends to produce is given by the right-hand motor rule.

The ampere is defined as the constant current which, flowing in two infinitely long straight parallel conductors of negligible cross-section placed in a vacuum one metre apart, produces between them a force of 2×10^{-7} newtons per metre of wire.

The rate of flow of charge is defined as the current.

$$\frac{Q}{t} = I$$

where Q is the quantity of charge
t is the time
I is the current

So that $Q = I \times t$

For a current of 1 ampere, the quantity of charge that flows in 1 second past a given point is defined as the coulomb, C.

$1\,C = 1\,A \times 1\,s$,   1 coulomb = 1 ampere-second

A component is said to have a resistance of 1 ohm when 1 volt is required to drive a current of 1 ampere through it.
$R = V/I$,   1 ohm = 1 volt per ampere, $1\,\Omega = 1\,V\,A^{-1}$
If Q coulombs of charge pass between two points A and B where the p.d. is V volts, the energy released W is given by

$W = Q\,V$ joules

If a current I passes for t seconds through a p.d. of V volts, the energy released is

$W = I\,t\,V$ joules

The e.m.f. (electromotive force) of a source is the energy converted when a coulomb of charge passes through the source. If 1 joule of energy is supplied by every 1 coulomb of charge, then the e.m.f. is 1 volt.

An electrical source is equivalent to an e.m.f. with a resistor in series, where the resistor corresponds to the internal resistance.

The e.m.f. of a source is equal to the p.d. across the terminals of the source when no current is being delivered: on open circuit.

The terminal potential difference of a source is related to its e.m.f. and internal resistance by the equation:

$$V = E - I \times r$$

where V = terminal potential difference
E = e.m.f. of source
r = internal resistance
I = current through the source

Electrical power P is given by

$$P = I \times V$$

1 watt = 1 ampere $\times$ 1 volt

Maximum power transfer occurs when the internal resistance of the source is equal to the external resistance of the circuit being supplied.

Problems

1   If a current of 40 mA passes through a lamp for 16 s, how much charge has passed any given point in the circuit?

2   A lightning flash lasted for 1 ms. If 5 C of charge was transferred during this time, what was the current?

3   The current in a circuit is 2.5×10^5 A. How long does it take for 500 C of charge to pass any given point in the circuit?

4   What is the p.d. across a 2 kΩ resistor if there is a current of 3 mA in it?

5   Draw a circuit diagram to show the experiment you would perform to determine the internal resistance of a battery. Explain what readings you would make and how you would calculate the internal resistance.

6   The terminal potential difference of a cell is 1.5 V on 'open circuit' and 1.4 V when connected in a circuit in which the current is 0.2 A. What is the internal resistance of the cell?

7   A 5 ohm resistor is connected to a 20 volt d.c. supply. Calculate the time for 3000 joules of energy to be converted in this resistor.

8   What size is the current in a circuit containing a battery of internal resistance 1.6 Ω and e.m.f. 4.2 V, connected to an external resistor of 2.6 Ω?

9   What is the terminal p.d. of a battery of internal resistance 1.8 Ω and e.m.f. 4.4 V when connected to an external resistor of 0.4 Ω?

10   What is the 'short-circuit' current for a battery of e.m.f. 4.5 V and internal resistance 0.9 Ω?

11   In order to determine the e.m.f. and internal resistance of a battery, a pupil used two 15 Ω resistors and an ammeter. When the two resistors were joined in parallel and connected to the battery the current was 2.0 A, and when connected in series the current was 0.75 A. Calculate the e.m.f. and internal resistance of the battery.

12   The circuit diagram shows two cells whose e.m.f.'s and internal resistances are known.

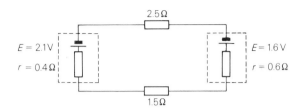

$E = 2.1\,\text{V}$
$r = 0.4\,\Omega$

$2.5\,\Omega$

$E = 1.6\,\text{V}$
$r = 0.6\,\Omega$

$1.5\,\Omega$

Calculate the current in the circuit.

13   Four identical cells each of e.m.f. 1.5 V and internal resistance 0.3 Ω are connected in series. What is the overall e.m.f. and internal resistance? What is the short-circuit current?

14   Six identical cells each of e.mf. 1.5 V and internal resistance 0.3 Ω are connected in parallel. What is the overall e.m.f. and internal resistance? What is the short-circuit current?

15   Calculate the power of a 120 ohm heater which operates on a 20 volt supply.

16   **a)**   What condition is necessary for a source with internal resistance to supply maximum power to an external circuit?
   b)   A car battery of internal resistance 0.05 Ω supplies power to a car. If the battery e.m.f. is 12.0 V, what is the maximum power which the battery can provide?

17   An electrical source with internal resistance r is used to operate a heater of resistance R. What fraction of the total power is available at the heater?

18   **a)**   A cell of e.m.f. E and internal resistance r is connected in series with an external resistor R.

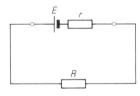

E r

R

For this circuit, show that $R = \dfrac{E}{I} - r$, where I is the current in the circuit.
   b)   The e.m.f. and internal resistance of a d.c. supply are to be measured. The d.c. supply is connected in series with an ammeter and an external resistor R of variable resistance.

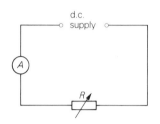

d.c. supply

A

R

The current I is measured for various values of R. The results are shown in the table.

R (ohms)	5.00	10.0	15.0	20.0	25.0
I (amps)	0.667	0.400	0.286	0.222	0.182

   i)   It would appear from the relationship in part (a) that a graph of R against $\dfrac{1}{I}$ will be a straight line. (E and r being taken as constant.) Using the data in the table, draw a graph to verify this statement. **(Use graph paper.)**
   ii)   **From this graph** calculate values for the e.m.f. and internal resistance of the d.c. supply. **Explain your working.**

SCEEB

19   **a)**   A d.c. supply has a constant e.m.f. of 12 V and an internal resistance of 3.0 Ω. A load resistor of resistance 1.0 Ω is connected across the supply terminals.
   Calculate   **i)**   the power delivered to the load resistor;
                 ii)   the voltage across the supply terminals.
   b)   Calculation of the power delivered to various values of load resistor is repeated giving the following results:

R in ohms	0.6	1.0	2.5	3.5	5.0	9.0
P in watts	6.7		11.9	11.9	11.3	9.0

   i)   Using suitable scales, draw a graph of P against R, including the point obtained in part (a) (i).
   ii)   From your graph determine the value of load resistor to which maximum power is delivered from this supply.
   iii)   Another 12 V d.c. supply has an internal resistance of 6.0 Ω. Suggest, with brief justification, the value of load resistor which should be connected to this supply for maximum power to be delivered to this load resistor.
   c)   On many signal generators there are two sets of output terminals – one set marked 6 Ω and the other marked 600 Ω. Suggest a reason for this provision.

SCEEB

4 d.c. circuits

4.1 Potential dividers

We now consider ways of varying the available potential difference. For this purpose we assume that the power supply has zero internal resistance and a fixed e.m.f.

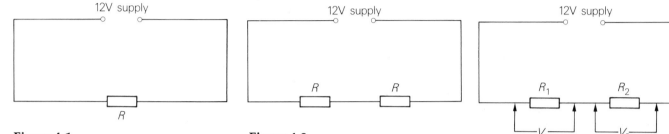

Figure 4.1   **Figure 4.2**

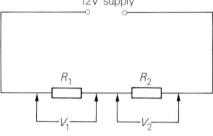

Figure 4.3

In Figure 4.1, the external resistor R is supplied with energy: 12 joules of energy for every coulomb of charge which passes through it ($12\,\text{J C}^{-1} \equiv 12\,\text{V}$).

If two identical external resistors are connected across the supply, Figure 4.2, the current through each resistor is the same: the rate of transformation of electrical energy to heat is the same for each resistor. The energy passing round the circuit is still $12\,\text{J C}^{-1}$, so 6 joules of energy is released for every coulomb of charge passing through each resistor. This means that the potential difference across each resistor is 6 volts, Figure 4.3.

$$V_1 + V_2 = 12$$

In general, for any number of resistors in series in a circuit, the total energy supplied by the source is the sum of the energies supplied to the resistors.

Figure 4.4 shows four unequal resistors in **series**. The total resistance in a series circuit is the sum of the resistances of the individual resistors.

$$R = R_1 + R_2 + R_3 + R_4$$

Because the resistors are in series, the current in each is the supply current I_s; multiply both sides of the above equation by I_s:

$$I_sR = I_sR_1 + I_sR_2 + I_sR_3 + I_sR_4$$

Then, because $V = I \times R$,

$$V = V_1 + V_2 + V_3 + V_4$$

The series of resistors in Figure 4.4 thus provide a range of potential differences.

The range of possible p.d.'s can be increased by using a continuous length of resistance wire. Figure 4.5 shows a source of e.m.f. of V_s volts and how various p.d.'s can be tapped off. The position of the contact point X on the resistance wire controls the p.d. V_{ox} between points O and X. Moving the contact from O to Y changes the p.d. V_{ox} as in the graph in Figure 4.6. The maximum value of V_{ox} is V_s when the contact is at point Y.

Halfway between O and Y, the p.d. is half V_s.

One-third of the way from O to Y the p.d. is one-third of V_s.

By tapping off various lengths of the resistance wire, smaller divisions of the total potential can be obtained. This arrangement is called a **potential divider**, the resistor OY being referred to as a **load resistor**.

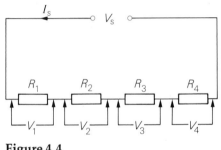

Figure 4.4

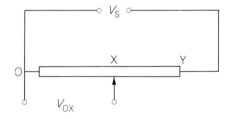

Figure 4.5 Potential divider.

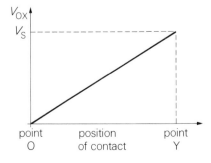

Figure 4.6 Variation of potential.

This useful arrangement can provide a completely variable range of p.d.'s by varying the position of a contact point on a variable resistor (rheostat) as shown in Figure 4.7.

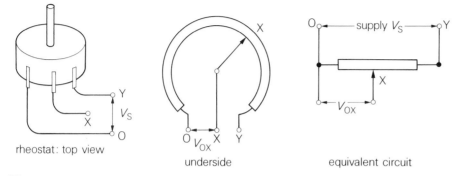

rheostat: top view   underside   equivalent circuit

Figure 4.7 Variable resistor as a potential divider.

Example 1

A potential divider consists of a rheostat of resistance 25 ohms and is connected across a supply. If a constant current of 60 milliamperes passes through the rheostat, what range of p.d.s can it provide?

With the contact at point O, the p.d. is zero,

i.e. $V_{ox} = 0$ because X and O coincide.

With the contact at point Y, the p.d. V_{oy} is V_s,

i.e. $V_{oy} = V_s = I_s R$

The total resistance R of the divider is 25 ohms
The current I_s through this resistor is 60×10^{-3} amperes

$\Rightarrow V_{oy} = 60 \times 10^{-3} \times 25 = 1.5$

$\Rightarrow$ maximum voltage obtainable is 1.5 V

The potential divider can provide a range of voltages from 0 V to 1.5 V

4.2 Resistors in parallel

In the circuit of Figure 4.8, the resistors R_1 and R_2 are connected in **parallel** to a supply of V_s volts giving a supply current of I_s amperes.

The system can be represented by an equivalent circuit shown in Figure 4.9 in which R_p is the equivalent resistance and I_s is the supply current.

A current I_s travels to a pair of parallel resistors and then separates at the junction into two currents I_1 and I_2, Figure 4.10.

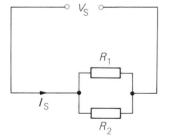

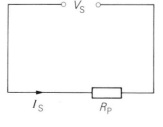

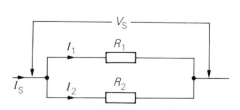

Figure 4.8   **Figure 4.9**   **Figure 4.10**

From Figure 4.10, $I_s = I_1 + I_2$

Also the two resistors are in parallel:

$\Rightarrow$ p.d. across R_1 = p.d. across $R_2 = V_s$

$\Rightarrow$ $I_1 \times R_1 = I_2 \times R_2 = V_s$

$\Rightarrow$ $I_1 = \dfrac{V_s}{R_1}$ and $I_2 = \dfrac{V_s}{R_2}$

From Figure 4.9, $I_s = \dfrac{V_s}{R_p}$

But $I_s = I_1 + I_2$

$\Rightarrow$ $\dfrac{V_s}{R_p} = \dfrac{V_s}{R_1} + \dfrac{V_s}{R_2}$

Divide both sides of this equation by V_s:

$$\frac{1}{R_p} = \frac{1}{R_1} + \frac{1}{R_2}$$

This important relationship gives the equivalent resistance of two resistors in parallel. In general, for more than two resistors in parallel, the equation becomes

$$\frac{1}{R_p} = \frac{1}{R_1} + \frac{1}{R_2} + \frac{1}{R_3} + \frac{1}{R_4} + \dots$$

Example 2

Calculate the equivalent resistance of four resistors of 2 ohms, 3 ohms, 4 ohms and 5 ohms connected in parallel

$$\frac{1}{R_p} = \frac{1}{2} + \frac{1}{3} + \frac{1}{4} + \frac{1}{5}$$

$$= \frac{30 + 20 + 15 + 12}{60} = \frac{77}{60}$$

$\Rightarrow R_p = \dfrac{60}{77}$ (remember to invert the fraction)

$\quad = 0.779$

The equivalent resistance of the four resistors is 0.78 ohms.

Example 3

Calculate the total resistance of the network in Figure 4.11.
    The circuit has a 'parallel' pair of resistors in series with the 2 ohm resistor, and all these are in parallel with a 5 ohm resistor.

a) The equivalent resistance R of the 'parallel' pair of resistors: 3 ohms and 4 ohms.

$$\frac{1}{R} = \frac{1}{3} + \frac{1}{4} = \frac{4 + 3}{12} = \frac{7}{12}$$

$\Rightarrow R = \dfrac{12}{7}$

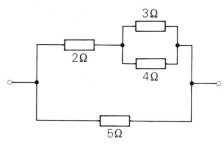

Figure 4.11

b) The equivalent resistance R' of the top arm of the circuit: 2 ohms and $\frac{12}{7}$ ohms. These are in series, so

$$R' = 2 + \frac{12}{7} = \frac{26}{7}$$

c) The equivalent resistance R'' of the whole circuit:
$\frac{26}{7}$ ohms and 5 ohms in parallel.

$$\frac{1}{R''} = \frac{1}{\frac{26}{7}} + \frac{1}{5} = \frac{7}{26} + \frac{1}{5}$$

$$\Rightarrow \frac{1}{R''} = \frac{35 + 26}{130} = \frac{61}{130}$$

$$\Rightarrow R'' = \frac{130}{61} \approx 2.1$$

The total resistance of the network is approximately 2 ohms.

4.3 Measuring resistance

We do not always know the resistance of a component in a circuit; in order to find it, we must measure both the current through the component and the potential difference across it. This gives the ratio V/I which is the resistance of the component.

However we must first consider in detail the use of meters in circuits.

Measurement of current

Consider the circuit in Figure 4.12 which contains a source of 2 volts, a resistor of 4 ohms and an ammeter to measure the current. The ammeter has a resistance of 1 ohm.

The total resistance of the circuit (resistor plus meter) is 5 ohms.

$$\Rightarrow \text{ the current } I = \frac{V}{R} = \frac{2}{5} = 0.4\,\text{A}$$

But an estimate of the current through a 4 ohm resistor connected to a 2 volt source gives

$$I = \frac{V}{R} = \frac{2}{4} = 0.5\,\text{A}$$

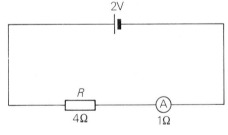

Figure 4.12

The ammeter in the circuit has increased the resistance of the circuit by 1 ohm, and so the measurement of the current is inaccurate. This may or may not be a serious problem: we shall see in the following two examples.

Example 4

An ammeter of resistance 5 ohms is used to measure the current in a circuit, Figure 4.13, consisting of a 12 volt source and a 5 ohm resistor. Calculate

a) the current expected with no ammeter in the circuit,
b) the current when the ammeter is in the circuit.

a) No ammeter

$$I = \frac{V}{R} = \frac{12}{5} = 2.4$$

The current is 2.4 A when there is no ammeter

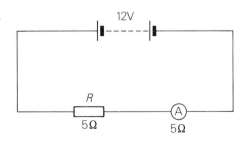

Figure 4.13

b) Ammeter included

Total resistance $R = 5 + 5 = 10$ ohms

$$I = \frac{V}{R} = \frac{12}{10} = 1.2$$

The current is 1.2 A when the ammeter is included.

Example 5

An ammeter of resistance 5 ohms is used to measure the current in a circuit, Figure 4.14, containing a 1000 ohm resistor and a 12 volt source. Calculate
a) the current expected with no ammeter in the circuit,
b) the current when the ammeter is in the circuit.

a) No ammeter

$$I = \frac{V}{R} = \frac{12}{1000} = 0.012$$

The current is 0.012 A when there is no ammeter

b) Ammeter included

Total resistance $R = 1000 + 5 = 1005$ ohms

$$I = \frac{V}{R} = \frac{12}{1005} = 0.012 \text{ (to 3 significant figures).}$$

The current is 0.012 A with the ammeter included

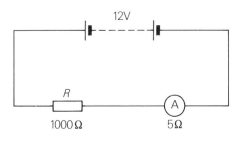

Figure 4.14

Thus, when the ammeter resistance is much **less** than that of the circuit components, there is less error. The real problem arises when attempting to measure the current in a circuit which has very low resistance. We can only attempt to reduce the error by making sure that the ammeter has as low a resistance as possible.

Measurement of potential difference

In the circuit in Figure 4.15, an attempt is made to measure the potential difference across a 1000 ohm resistor which is connected to another 1000 ohm resistor and a 2 volt source. The voltmeter also has a resistance of 1000 ohms.
a) No voltmeter in the circuit

total resistance $R = 1000 + 1000 = 2000 \ \Omega$

supply voltage $V = 2 \text{ V}$

$\Rightarrow$ current $I = \frac{V}{R} = \frac{2}{2000}$

The p.d. across AB, $V_{AB} = I \times R = \frac{2}{2000} \times 1000$

$\Rightarrow$ $V_{AB} = 1.0 \text{ V}$

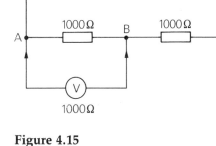

Figure 4.15

With no voltmeter in the circuit, we would expect a p.d. of 1 V across the 1000 ohm resistor.

b) Voltmeter included in circuit
The circuit is now different because the total resistance is reduced. The resistance of the voltmeter is in parallel with that of the 1000 ohm resistor:

$$\frac{1}{R_{AB}} = \frac{1}{1000} + \frac{1}{1000} = \frac{2}{1000}$$

$$\Rightarrow R_{AB} = \frac{1000}{2} = 500 \ \Omega$$

The total resistance of the circuit is now $(500 + 1000)$ ohms $= 1500$ ohms

$$\Rightarrow \qquad \text{current } I = \frac{V}{R} = \frac{2}{1500}$$

The p.d. acros AB, $\quad V_{AB} = I \times R = \frac{2}{1500} \times 500$

$$\Rightarrow \qquad V_{AB} = 0.66 \text{ V}$$

When the voltmeter is in the circuit, the p.d. is 0.66 V across the 1000 ohm resistor.

If a circuit contains resistors with values near to the resistance of the voltmeter, the measurement of p.d. is unreliable.

Example 6

A voltmeter of resistance 1000 ohms is used to measure the p.d. across a 20 ohm resistor AB in a circuit, Figure 4.16, containing another resistor of 1180 ohms and a 12 V source. Calculate
a) the p.d. with no voltmeter in the circuit,
b) the p.d. when the voltmeter is in the circuit.

a) No voltmeter in the circuit
Total resistance $R = 20 + 1180 = 1200 \, \Omega$

$$\Rightarrow I = \frac{V}{R} = \frac{12}{1200} = 0.01 \text{ A}$$

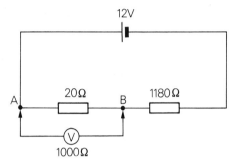

Figure 4.16

The p.d. across AB, $V_{AB} = IR_{AB} = 0.01 \times 20 = 0.2 \, V$

The p.d. across AB is 0.2 volts

b) Voltmeter included in circuit
The resistance of the voltmeter is now in parallel with the 20 ohm resistor, so

$$\frac{1}{R_{AB}} = \frac{1}{20} + \frac{1}{1000} = \frac{50 + 1}{1000} = \frac{51}{1000}$$

$$\Rightarrow R_{AB} = \frac{1000}{51} = 19.6$$

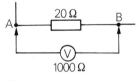

Figure 4.17

The total resistance of the circuit is now $(19.6 + 1180)$ ohms $= 1199.6$ ohms

$$\Rightarrow \text{ current } I = \frac{V}{R} = \frac{12}{1199.6} = 0.01 \text{ A}$$

The p.d. across AB, $V_{AB} = IR_{AB} = 0.01 \times 19.6 = 0.196 \text{ V}$

The p.d. across AB is approximately 0.2 volts.
(The voltmeter is unlikely to show the small difference in the reading).

Thus, when the voltmeter resistance is much **greater** than that of the circuit component, there is less error because the current in the circuit is hardly affected and the p.d. across the component then remains practically the same. A good voltmeter should have a very high resistance.

Measurement of resistance using voltmeter and ammeter

In order to measure resistance, meters are needed for both the p.d. across the resistor and the current through it. There are two possible ways of connecting the meters, shown below.

The circuit in Figure 4.18 is suitable for the measurement of low values of

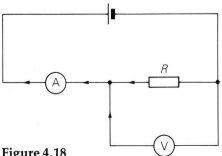

Figure 4.18

resistance. The voltmeter measures the p.d. across R and the ammeter measures the sum of the currents through R and the voltmeter. The voltmeter has a much larger resistance than R and the current through it is very much smaller than that through R so that the error in determining R can be neglected.

The circuit in Figure 4.19 is suitable for the measurement of high values of resistance. The voltmeter measures the p.d. across R and the ammeter, but the ammeter resistance is usually so small in comparison that the p.d. across it makes little difference to the accuracy of the result for R. The ammeter measures the true current through R using this circuit.

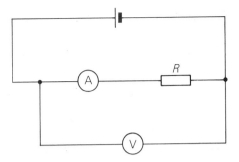

Figure 4.19

4.4   Wheatstone bridge circuit

The Wheatstone bridge circuit, a resistor network for determining resistance was devised by the English physicist Sir Charles Wheatstone. It can provide more accurate measurements of resistance than the voltmeter-ammeter method and in fact does not depend for its accuracy on the accuracy of a meter. It does however require a **sensitive** meter. The form of circuit most often encountered is shown in Figure 4.20. G represents a sensitive centre-zero galvanometer.

It is often helpful to consider this circuit re-drawn as in Figure 4.21. (The galvanometer has been omitted.)

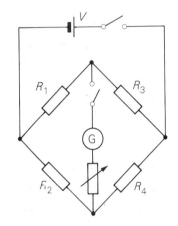

Figure 4.20 Wheatstone bridge circuit

Consider AC as a potential divider made from two resistors R_1 and R_3. Applying Ohm's Law:

  p.d. across R_1 $V_{AB} = I_1 R_1$

  p.d. across R_3 $V_{BC} = I_1 R_3$

        $V = V_{AB} + V_{BC}$

Consider DF as another potential divider in parallel with the first:

  p.d. across R_2 $V_{DE} = I_2 R_2$

  p.d. across R_4 $V_{EF} = I_2 R_4$

        $V = V_{DE} + V_{EF}$

If the resistor values are chosen so that $V_{AB} = V_{DE}$ then it follows also that $V_{BC} = V_{EF}$ since the sum of the individual p.d.'s gives the supply p.d. which is V.

$$\Rightarrow I_1 R_1 = I_2 R_2 \quad \Rightarrow \frac{R_1}{R_2} = \frac{I_2}{I_1}$$

and $I_1 R_3 = I_2 R_4 \quad \Rightarrow \frac{R_3}{R_4} = \frac{I_2}{I_1}$

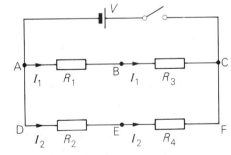

Figure 4.21

so that $\dfrac{R_1}{R_2} = \dfrac{R_3}{R_4}$ or $\dfrac{R_1}{R_3} = \dfrac{R_2}{R_4}$

Now when $V_{AB} = V_{DE}$, then a galvanometer connected between points B and E registers zero current (null deflection) since points B and E are at the same potential: there is no potential difference between them and so $V_{BE} = 0$.

When the resistor values are selected so that the meter reads zero current, the bridge is said to be **balanced**.

In practice three of the resistors in this circuit are accurately known standard resistors while the fourth one is of unknown value to be determined using the bridge. By choosing a suitable ratio for resistors R_1 and R_3 and then gradually adjusting the value of R_2 until the bridge is balanced, it is possible to find R_4 from the ratio

$$\frac{R_1}{R_3} = \frac{R_2}{R_4}$$

The size of the current in the galvanometer depends on the extent to which the resistor values are not in the correct ratio. A protective variable resistor is inserted in series with the sensitive galvanometer in order to minimize the risk of damage when the 'out of balance' current is high, Figure 4.20.

Example 7

The Wheatstone bridge shown in Figure 4.22 is balanced. If $R_1 = 120\ \Omega$, $R_2 = 400\ \Omega$ and $R_3 = 80\ \Omega$, what is the value of the unknown resistor X?

$$\text{Since the bridge is balanced } \frac{R_1}{R_2} = \frac{R_3}{X}$$

$$\Rightarrow \qquad X = \frac{R_2 \times R_3}{R_1}$$

$$\Rightarrow \qquad X = \frac{400 \times 80}{120} = 267$$

The value of the unknown resistor X is 267 ohms

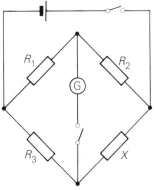

Figure 4.22

The unbalanced Wheatstone bridge

The circuit in Figure 4.23 represents a balanced bridge.

If the value of resistor R_1 is increased by a small amount ΔR, then the galvanometer indicates an out-of-balance current I. The set of results in Table 1 illustrates typical behaviour of this circuit.

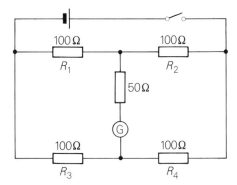

Figure 4.23

Increase in resistance $\Delta R(\Omega)$	0	1.0	2.0	3.0	4.0	5.0	6.0	7.0	8.0	9.0	10
Out of balance current I (µA)	0	35	70	100	130	170	200	230	270	300	340

Table 1

A graph of change in resistance ΔR against current I indicates that, for an initially balanced bridge, as the value of one resistor is changed by a small amount, the current is proportional to the change in resistance, Figure 4.24.

If resistor R_1 is replaced by a heat sensitive resistor called a thermistor then the out-of-balance current indicates temperature change, Figure 4.25. Alternatively, resistors R_1 and R_2 could be replaced by an identical pair of electrical strain gauges (resistors whose resistance alters with changes in stress) and the out-of-balance current would give information about mechanical strain, Figure 4.26. There are many applications of the Wheatstone bridge circuit.

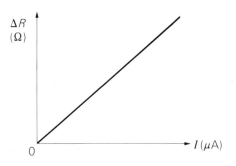

Figure 4.24

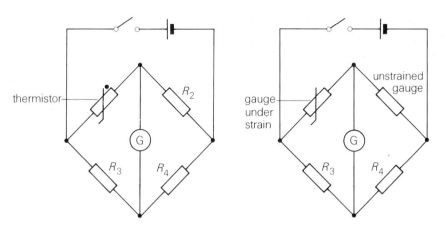

Figure 4.25 Measuring temperature.   **Figure 4.26** Measuring strain.

Metre bridge

An alternative form of Wheatstone bridge circuit which uses only one standard resistor is called the metre bridge, Figure 4.27. With this arrangement two of the resistors are replaced by a one-metre length of uniform resistance wire with a movable contact P.

The position of the contact point P decides the ratio R_1 to R_2 since, provided the wire is of uniform thickness, the ratio can be expressed as the ratio of the two lengths of wire l_1 to l_2

i.e. $$\frac{R_1}{R_2} = \frac{l_1}{l_2} = \frac{R_3}{R_4}$$

and since the wire is 1 metre long, $l_2 = 100 - l_1$ the ratio is

$$\Rightarrow \quad \frac{l_1}{100 - l_1} = \frac{R_3}{R_4}$$

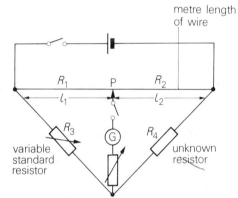

Figure 4.27 Metre bridge

This method provides an accurate determination of resistance: the error in the measurement of l_1 and l_2 can be less than 1% since these lengths can be measured to the nearest millimetre.

Example 8

A metre bridge is balanced when the movable contact is 64 cm from the end of the wire, Figure 4.28. What is the value of the unknown resistor?

At balance $\dfrac{l_1}{100 - l_1} = \dfrac{R_3}{R_4}$ $l_1 = 64\,\text{cm}$ $R_3 = 100\,\Omega$ $R_4 = X$

$$\Rightarrow \quad \frac{64}{100 - 64} = \frac{100}{X}$$

$$\Rightarrow \quad \frac{64}{36} = \frac{100}{X}$$

$$\Rightarrow \quad X = \frac{36 \times 100}{64}$$

$$X = 56.25$$

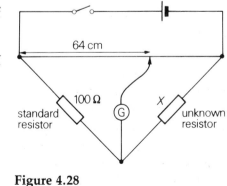

Figure 4.28

The unknown resistor has a value of 56 ohms

4.5   Design and uses of resistors

Table 2 provides details of construction, accuracy and power handling of the three main types of resistor in use.

Type	Construction	Accuracy	Power handling
wirewound	alloy wire on ceramic rod	±0.1%	2–25 W
thin film	metal oxide on ceramic rod	±2%	0.1–0.5 W
carbon composition	carbon-clay mixture	±10%	0.1–2.0 W

Table 2

The standard resistors used in the Wheatstone bridge circuit are wirewound because of their great accuracy.

Thin film resistors are not as temperature dependent as carbon composition resistors and are gradually replacing them as they become cheaper.

We have described the use of resistors to protect galvanometers by reducing the possible current and also as load resistors for potential dividers, but a resistor may be used merely as a means of converting electrical energy to heat or light energy. An electric heater is an example of a resistor specifically for this purpose.

Summary

A potential divider may be used to provide fixed or variable p.d.'s from a given source, using fixed or variable resistors.

The total resistance in a series circuit is given by the sum of the individual resistors:

$$R_s = R_1 + R_2 + R_3 + \dots$$

For resistors in parallel the total resistance can be calculated from:

$$1/R_p = 1/R_1 + 1/R_2 + 1/R_3 + \dots$$

An ammeter is connected in series in a circuit. A good ammeter has a low resistance.

A voltmeter is connected in parallel in a circuit. A good voltmeter has a high resistance.

For a balanced Wheatstone bridge $\dfrac{R_1}{R_2} = \dfrac{R_3}{R_4}$

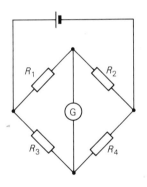

For an initially balanced bridge, as the value of one resistor is changed by a small amount, the galvanometer current is proportional to the change in resistance.

For a balanced metre bridge

$$\frac{l_1}{100 - l_1} = \frac{R_3}{R_4}$$

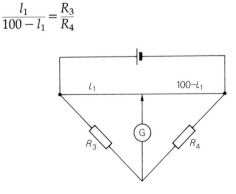

Problems

1 The following series of resistors is connected to a 12 V supply.

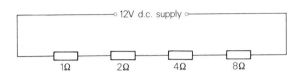

a) Calculate the p.d. across each resistor.
b) List all the possible p.d.'s obtainable from this resistor arrangement.

2 A potential divider has resistance 10 ohms. If a constant current of 300 mA flows through the divider what range of voltages can it provide?

3 Derive an expression for the combined resistance of two resistors in parallel.

4 Calculate the net resistance in the following circuit.

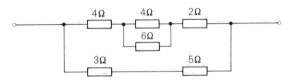

5 Derive an expression for the combination of two resistors in series.

6 An ammeter of resistance 10 ohms is used to measure the current in the circuit shown.

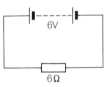

By how much does the measured current differ from the expected current?

7 A voltmeter of resistance 1000 ohms is used to measure the p.d. across one of the resistors in the circuit shown.

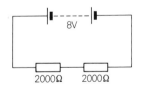

By how much does the measured p.d. differ from the expected p.d.?

8 A student attempts to find experimentally the value of two resistors connected in series. The resistors have values 4 ohms and 2 kilohms. If the student has available an ammeter of resistance 5 ohms and a voltmeter of resistance 1000 ohms, what measurement should be made to obtain the most accurate results?

9 Calculate the total equivalent resistance between points X and Y in the circuit shown.

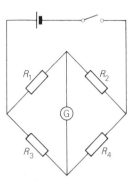

10 The circuit shown represents a Wheatstone bridge circuit.

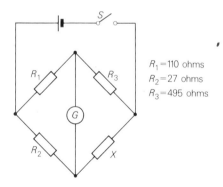

Explain the use of this circuit and derive an expression for the balance condition.

11 The Wheatstone bridge circuit shown below is balanced.

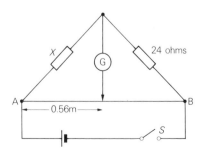

$R_1 = 110$ ohms
$R_2 = 27$ ohms
$R_3 = 495$ ohms

Calculate the value of the unknown resistor X.

12 Describe how the out-of-balance current in a Wheatstone bridge circuit can be used to measure temperature.

13 A metre bridge is balanced when the movable contact is 0.56 m from end A of the wire as shown.

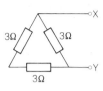

What is the value of the unknown resistor X?

14 a) Describe an experiment to investigate how the rate at which heat is produced in a coil of wire varies with the current in the wire. Include in your answer:
  i) a labelled diagram of the apparatus used;
  ii) a description of the experimental measurements made;
  iii) details of any measures taken to minimise experimental error.
b) In such an experiment the following results were obtained.

Rate of heat production (J s⁻¹)	1.9	3.2	11	16	24
Current in wire (A)	0.7	0.9	1.6	2.0	2.4

  i) Use these results to determine graphically the relationship between the rate of production of heat and the current in the wire.
  ii) From the graph determine the resistance of the wire.

SCEEB

15 To find an unknown resistance R, a pupil sets up the following circuit where the voltmeter can be connected across either XY or XZ.

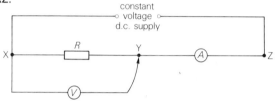

The results obtained are shown in the table.

Voltmeter connected across	Voltmeter reading V	Ammeter reading mA
XY	1.00	50
XZ	1.10	50

a) The current is the same for both positions of the voltmeter. From this information, what can you deduce about the voltmeter?
b) Find the resistance R and the resistance of the ammeter.
c) If the voltmeter is disconnected and another voltmeter of resistance 180 Ω is connected across XY, what would be the readings on the ammeter and this new voltmeter?   *SCEEB*

16 Instead of using a transformer, bulbs may be connected in series with a 'mains dropping' resistor R across the 240 V supply as shown.

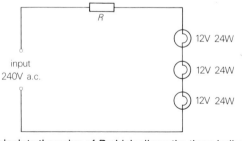

a) Calculate the value of R which allows the three bulbs to be run at their correct brightness.
b) What fraction of the input power is used by the bulbs?
c) Give the reason why the power used by the bulbs is less than the input power.   *SCEEB*

17 The diagram shows how a resistor is used to convert an ammeter of resistance 1.0 Ω and full scale deflection of 5.0 mA to one which has a full scale deflection of 500 mA.

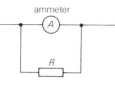

a) What resistance R is required?

b) State how the 5 mA meter can be converted into a voltmeter with a full scale deflection of 3 V. Calculate the resistance of any resistor required.

c) In the circuit below, the 2 V cell has negligible internal resistance.

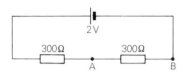

i) What is the potential difference across AB?

ii) The voltmeter in (b) is connected across AB. What is the reading on this voltmeter?

iii) Account for the difference in your answers to (i) and (ii).

SCEEB

18 The diagram shows a metre bridge circuit used to find the resistance of X.

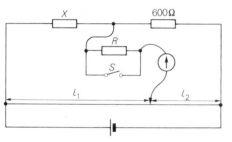

a) Explain the functions of resistor R and switch S.

b) Explain why a balance point in the middle section of the bridge wire gives a more accurate value of X.

c) If at balance point $l_1 = 60 \pm 1$ cm and $l_2 = 40 \pm 1$ cm, calculate the resistance of X and give the error limits of your answer.

d) Indicate two sources which contribute to the error in the measurements l_1 and l_2.

SCEEB

19 Two pupils set out to find the resistance of an unknown resistor X.

a) Pupil A sets up a metre bridge circuit using resistance box R, the resistance of which may be varied from 1 Ω to 10 000 Ω.

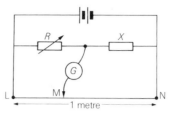

Balance point is found when $R = 3000$ Ω and LM = 107 mm.

i) Calculate the resistance of X.

ii) Suggest any change pupil A could make to increase the accuracy of the result.

b) Pupil B sets up an ammeter-voltmeter circuit.

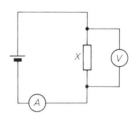

The reading on the ammeter is 90 μA and the reading on the voltmeter is 1.5 V.

i) Calculate the resistance of X.

ii) Suggest any change pupil B could make to increase the accuracy of the result.

c) Discuss whether method (a) or method (b) is more accurate.

SCEEB

5 Electric field

5.1 Introduction

A polythene rod rubbed with a duster becomes negatively charged as a result of friction: the rod gains electrons from the duster. Similarly, a nylon rod can be positively charged by friction because it loses electrons to the duster.

There are only two types of electric charge: positive and negative. A pair of charges gives rise to an **electric force** between them. If two rods that have been oppositely charged are brought near to each other, there is a force of attraction between them. Two rods having the same type of charge repel each other.

The movement of air during a thunderstorm causes thunder clouds to be charged by friction. Very large quantities of positive and negative charge are built up: when these are discharged, a huge spark of lightning results. Figure 5.1.

A flow of electron current occurs when an electric fire is switched on: electrical energy is converted to heat energy and light energy.

There are two ways of detecting charge:
a) by measuring the **flow** of charge, e.g. electric current,
b) by measuring the **force** between charges, e.g. the charged polythene and nylon rods.

Figure 5.1 Lightning

5.2 Electric field

The gravitational force of attraction of the mass of the Earth for other masses can be described in terms of a gravitational field. The force between charged particles can similarly be described in terms of an **electric field**.

The fundamental unit of charge is the charge on an electron and the charge on a proton. However, in describing electric forces, we shall be concerned only with the effects of groups of these charged particles which are spread over the surface of an object: these are simply described as the **charge** on an object.

In Figure 5.2, the two small objects Q_1 and Q_2 are identical and both are negatively charged. They are placed some distance apart, but there is a force of repulsion between them because they both have a negative charge.

Figure 5.3 shows a small charged object Q_3 and a large object Q_4 which is also charged. The charge on Q_4 is spread over a larger surface so that the forces of repulsion are more complicated: the forces are in many directions and the electrons are at various distances from each other.

To simplify the problem, we must consider only very small charged objects called **point charges**.

force
(a)
force

Figure 5.2

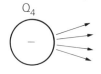

(b)

Figure 5.3

Direction of electric field force

In Figure 5.2, the force exerted by charge Q_1 on charge Q_2 is directed from Q_1 to Q_2 along the line joining them. There is an equal electric force in the opposite direction from Q_2 to Q_1.

The apparatus shown in Figure 5.4 is used to investigate the electric field. The glass dish contains oil on which is sprinkled a little grass seed. Two metal 'point' electrodes dip into the oil, and they are connected to a high voltage

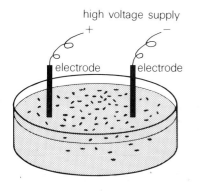
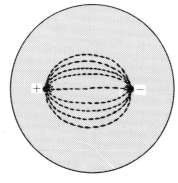

high voltage supply

electrode electrode

dish containing oil and grass seed

pattern of seeds in oil

Figure 5.4

supply. When the supply is switched on, the electrodes become positively and negatively charged and the seeds form a pattern as shown. The lines in the pattern indicate the direction of the electric force, and the pattern shows the electric field between the two electrodes.

Distance and electric field force

There is a relationship between the force F exerted between two identical charged objects and the distance d between them. This was first investigated by the French scientist C.A. Coulomb in 1785. He used an instrument called a torsion balance, Figure 5.5.

A simpler method uses a small lightweight sphere that is coated with metal and suspended by means of a long non-conducting thread in front of a lamp, Figure 5.6. The sphere is charged and repelled by an identically charged sphere on an insulating rod that is brought near to it.

The two spheres cast shadows on a screen from which the relative distances can be measured. Measurement of the actual distances between the spheres would involve the danger of changing the charge on them. The distance d_1 on the screen represents the distance between the two spheres; d_2 is the displacement of the suspended sphere and is therefore a measure of the electric force F between the two spheres.

Values of d_2 corresponding to F are recorded for various values of d_1. The resulting graph is shown in Figure 5.7. The results can be shown to give a straight line through the origin if F is plotted against $1/d_1^2$, Figure 5.8.

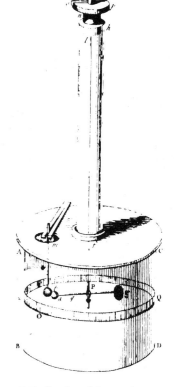

Figure 5.5 Coulomb's torsion balance

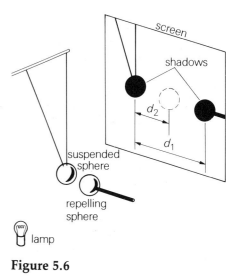

Figure 5.6

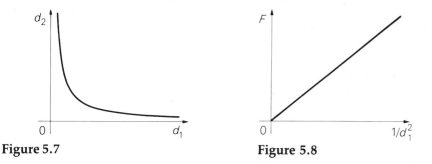

Figure 5.7

Figure 5.8

Figure 5.8 shows that $F \propto 1/d_1^2$, since d_2 is a measure of the electric force F.

The electric field force F is inversely proportional to the square of the distance d_1^2 between the two charged spheres.

5 Electric field

Charge and electric field force

The same equipment can be used to measure the electric force F (i.e. d_2) between two identical but differently charged spheres.

The suspended sphere is first charged by contact with a charged rod. The repelling sphere is then given an identical charge Q by momentary contact with the suspended sphere, Figure 5.9, because the charge is shared equally between the two identical spheres.

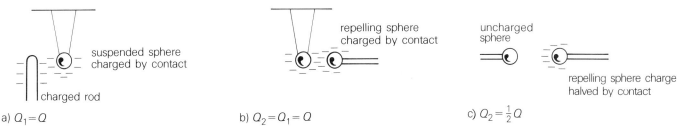

a) $Q_1 = Q$ b) $Q_2 = Q_1 = Q$ c) $Q_2 = \frac{1}{2}Q$

Figure 5.9

The charge on the repelling sphere can be successively halved by momentary contact with an identical uncharged sphere. In this way, a succession of values of d_2 are obtained for constant d_1 in which the suspended sphere charge remains Q and the charge on the repelling sphere is Q, $Q/2$, $Q/4$, $Q/8$, etc.

Figure 5.10 shows the results of the experiment; if the product of the charges Q_1 and Q_2 is plotted against the electric force F (i.e. d_2) for two spheres placed a constant initial distance apart, a straight line graph is obtained.

$$F \propto Q_1 Q_2$$

The electric field force F between two charges

Q_1 and Q_2 is directly proportional to the product

of the charges.

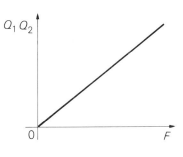

Coulomb's Law

The results of the above two experiments can be combined in Coulomb's Law:

Figure 5.10

$$F \propto \frac{Q_1 Q_2}{d^2}$$

The force F between two small charged objects

placed a distance d apart varies

       as the product of the magnitude of the charges

and,   as the inverse square of the distance between

       the charges.

$$\Rightarrow F = \frac{k\, Q_1\, Q_2}{d^2} \text{ where } k \text{ is a constant.}$$

The constant k depends on the properties of the medium through which the charges exert their forces.

Example 1

Using Coulomb's Law, calculate the force between a negative charge of 4 microcoulombs and a positive charge of 6 microcoulombs when placed 0.12 m apart. (The value for k in the equation should be taken as 9×10^9.)

Coulomb's Law: $F = \dfrac{kQ_1 Q_2}{d^2}$ where $\quad k = 9 \times 10^9$

$$Q_1 = -4 \times 10^{-6}\,\text{C}$$

$$Q_2 = 6 \times 10^{-6}\,\text{C}$$

$$d = 0.12\,\text{m}$$

$\Rightarrow \qquad\qquad F = \dfrac{9 \times 10^9 \times 4 \times 10^{-6} \times 6 \times 10^{-6}}{0.12 \times 0.12}$

$\Rightarrow \qquad\qquad F = 15\,\text{N}$

The charges exert a force of 15 N on each other

5.3   Field patterns and lines of force

The field strength g of the Earth's gravitation has an approximate value of $10\,\text{N kg}^{-1}$. This means that the gravitational force of attraction F_g of the mass of the Earth is mg newtons on a mass of m kg near the Earth.

$$F_g = mg$$

Electric field strength E is the electrical force on a unit charge in the electric field and has units of newtons per coulomb (N C^{-1}). Thus the electric force F_E on a charge of Q coulomb in this field is QE newtons.

$$F_E = QE$$

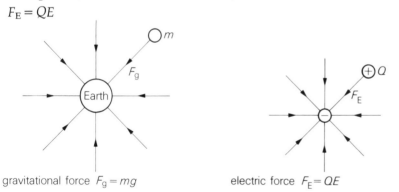

gravitational force $F_g = mg$ electric force $F_E = QE$

Figure 5.12

Figure 5.11 Gravitational and electric field patterns

The two types of field with their lines of force are shown in Figure 5.11. The arrows on the lines of force indicate the direction in which the mass or **positive** charge moves. If the object creating the electric field force is positively charged, then the arrows on the lines of force are reversed, Figure 5.12, so that they still indicate the direction of movement of a positive charge.

An electric field pattern may be plotted from the shadows cast on a screen from a lamp, Figure 5.13. The direction of each line of force is found from the shadow cast by a thin metal foil that is charged and held on an insulating rod between the charged metal plates. Each line of force can be marked on the screen. Figure 5.14 shows the electric field patterns obtained in this way for various arrangements.

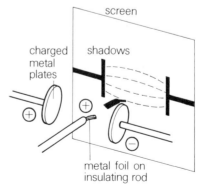

Figure 5.13

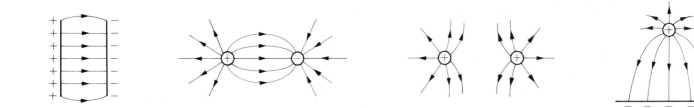

Figure 5.14 Electric field patterns

The variation in strength of the electric field is indicated by the spacing of the lines of force; the closer the spacing, the stronger is the field. Notice that the lines of force between a pair of parallel charged metal plates are evenly spaced showing that the electric field is uniform.

In calculations concerning a charge in an electric field, we shall always assume that the field is uniform.

Example 2

The charge on an electron is -1.6×10^{-19} C. Calculate the electric force on an electron in a uniform electric field of 1000 N C^{-1}.

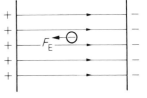

$Q = -1.6 \times 10^{-19}$ C; $E = 1000$ N C^{-1}

electric force $F_E = Q \times E$

$\Rightarrow \qquad F_E = -1.6 \times 10^{-19} \times 1000$

$\Rightarrow \qquad F_E = -1.6 \times 10^{-16}$

The force on the electron is -1.6×10^{-16} N

Note that the force is negative: this shows that the force acts in a direction opposite to that of the electric field.

Example 3

A charged oil drop has a mass of 9.6×10^{-15} kg. If the drop is stationary between horizontal charged parallel plates in an electric field of strength 2×10^5 N C^{-1}, calculate the charge on the oil drop.

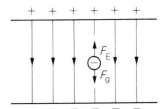

Here the electric force and the gravitational force on the oil drop are balanced: the electric force on the oil drop must be upwards.

downwards gravitational force $F_g = mg$

$\Rightarrow F_g = 9.6 \times 10^{-15} \times 10$

$\Rightarrow F_g = 9.6 \times 10^{-14}$ N

upwards electric force $F_E = QE$

$\Rightarrow F_E = Q \times 2 \times 10^5$ N

As the two forces are balanced, $F_g = F_e$

$\Rightarrow 9.6 \times 10^{-14} = Q \times 2 \times 10^5$

$\Rightarrow \qquad Q = \dfrac{9.6 \times 10^{-14}}{2 \times 10^5}$

$\Rightarrow \qquad Q = 4.8 \times 10^{-19}$

The oil drop has a charge of 4.8×10^{-19} C

5.4   Field strength and potential difference

If the lines of force in an electric field are evenly spaced, the field is uniform. This means that a charged particle experiences the same force throughout the electric field.

A positive charge Q is placed in a uniform electric field of strength E between two metal plates that are placed a distance d apart, Figure 5.15. The electric field causes the charge to move from the positive to the negative plate. We can find the work done by the field on the charge.

work done = force × displacement

$$W = F_E \times d$$
$$= QEd$$

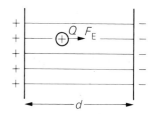

Figure 5.15

In section 3.2, the potential difference V between two points was defined as the work done W in moving a charge Q between the two points.

$$W = Q V$$

$$\Rightarrow Q Ed = Q V$$

$$\Rightarrow \quad E = \frac{V}{d}$$

Thus the field strength E newtons per coulomb is equivalent to the potential difference per unit distance V/d volts per metre, sometimes called **potential gradient**.

$$N\,C^{-1} \equiv V\,m^{-1}$$

Electric field strength may be measured in $N\,C^{-1}$ or in $V\,m^{-1}$.

Example 4

A potential difference of 2kV is applied across a pair of parallel metal plates 10 cm apart. What is the electric field strength between them?

$$V = 2 \times 10^3\,V; \qquad d = 10 \times 10^{-2}\,m$$

$$E = \frac{V}{d}$$

$$\Rightarrow E = \frac{2 \times 10^3}{10 \times 10^{-2}} = 2 \times 10^4$$

The electric field strength is 2×10^4 V m^{-1}

Example 5

Calculate the potential difference required to balance an oil drop of mass 2×10^{-15} kg and charge -8.0×10^{-19} C between two horizontal metal plates 5 cm apart.

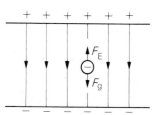

Because the forces on the drop are balanced, $F_E = F_g$

$$\Rightarrow Q E = mg \qquad \text{where } m = 2 \times 10^{-15}\,kg \qquad Q = -8 \times 10^{-19}\,C$$

$$\Rightarrow Q \frac{V}{d} = mg \qquad\qquad g = 10 \text{ m s}^{-2} \qquad d = 5 \times 10^{-2}\,m$$

$$\Rightarrow -8 \times 10^{-19} \times \frac{V}{5 \times 10^{-2}} = 2 \times 10^{-15} \times 10$$

$$\Rightarrow V = \frac{2 \times 10^{-15} \times 10 \times 5 \times 10^{-2}}{-8 \times 10^{-19}} = -1250$$

A potential difference of 1250 V is needed to balance the drop

Example 6

The charge on an electron is -1.6×10^{-19}C. Calculate the increase in energy if an electron is accelerated from rest through a distance of 2 cm where the uniform electric field strength is 1000 N C^{-1}.

$Q = -1.6 \times 10^{-19}$C; $E = 1000$ N C^{-1}; $d = 0.02$ m

Work done by field $W = Q V$

$$E = \frac{V}{d} \Rightarrow V = Ed$$

$$\Rightarrow W = Q\,Ed$$

$$\Rightarrow W = -1.6 \times 10^{-19} \times 1000 \times 0.02 = -3.2 \times 10^{-18}$$

The electron gains 3.2×10^{-18} joules

5.5   Movement of charged particles in electric fields

The electron beam in the tube of a cathode ray oscilloscope moves parallel to an electric field in the electron gun and at right angles to an electric field between the Y-plates, Figure 5.16.

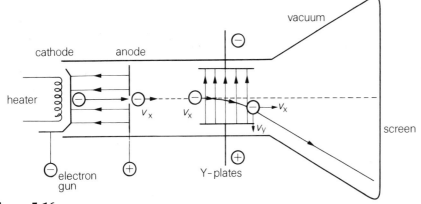

Figure 5.16

Electrons released from the heated cathode are accelerated by the uniform electric field between the cathode and anode. On reaching the anode they travel at a constant velocity v_x until they reach the region of electric field between the Y-plates. The passage of electrons through this region, where the field is at right angles to their motion, resembles the trajectory of a projectile when projected horizontally through the Earth's gravitational field, and in fact the electrons follow a similar parabolic path. The horizontal velocity v_x of the electrons remains constant as they travel from the electron gun through a vacuum towards the Y-plate region. In this region there is an acceleration towards the lower, positive plate and they gain a vertical component of velocity v_y as well as maintaining their horizontal velocity v_x. The electron beam strikes the screen, having been deflected by the electric field. The electric field between the Y-plates depends on the p.d. applied to them since $E = V/d$ and so the deflection of the beam can be used to compare p.d.'s. In a solid the application of an electric field to a conducting material causes the 'free electrons' to drift. In a circuit, the applied potential difference sets up an electric field within the conductors, giving the conduction electrons their drift velocity.

Summary

The force between two small charged objects varies directly as the sizes of the charges and inversely as the square of the distance between them.

The direction of an electric field is given by the direction of the force on a positive charge placed in the field.

The magnitude of an electric field is given by the magnitude of the force on a unit positive charge placed in the field and is measured in newtons per coulomb, $N\,C^{-1}$.

The electrical force on a charge of Q coulombs in an electric field of strength E newtons per coulomb is $Q \times E$ newtons.

$$F_E = Q \times E$$

The work done W in moving a charge Q a distance d against an electric field of strength E is given by

$$W = Q \times E \times d \text{ joules}$$

The work done per coulomb of charge is $E \times d$ joules

The potential difference between two points is equal to the work done per unit charge in moving charge from one point to another.

$$V = E \times d$$
$$\Rightarrow E = V/d \text{ volts per metre}$$

The work done W in transferring Q coulombs of charge through a p.d. of V volts is given by $W = Q \times V$ joules.

Problems

1   What does the force between two small charged objects depend on?

2   State Coulomb's Law.

3   Draw the electric field pattern near to a positively charged sphere. State the unit of electric field strength.

4   Calculate the electric force on an electron in a uniform electric field of $1200\ N\,C^{-1}$ (electron charge $= -1.6 \times 10^{-19}C$)

5   A charge of $3\ \mu C$ experiences a force of $2 \times 10^{-2}\ N$ in an electric field. What is the value of the electric field strength?

6   An electric field of $5 \times 10^5\ N\,C^{-1}$ exerts a force of $2 \times 10^{-2}\ N$ on a charged object. What is the magnitude of the charge on the object?

7   A charged oil drop of mass 8.6×10^{-15} kg is held stationary in a uniform electric field. If the drop has a charge of $4.8 \times 10^{-19}C$, what is the electric field strength?

8   The strength of a uniform electric field between two metal plates separated by 0.12 m is $3 \times 10^4\ V\,m^{-1}$. What is the potential difference across the pair of plates?

9   A p.d. of 1400 volts is required to balance an oil drop of mass 3×10^{-15} kg between two metal plates 4 cm apart. What is the charge on the oil drop?

10   The electrons in a cathode ray tube beam are accelerated from cathode to anode by a p.d. of 2500 V. If this p.d. is increased to 10 000 V, how many times greater will a) the kinetic energy and b) the velocity of the electrons be?

11   An electron enters a uniform electric field at right angles to the field direction as shown.

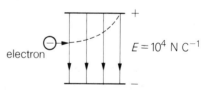

If the electric field strength is $10^4\ N\,C^{-1}$, what is the electron's acceleration towards the positive plate? (mass of electron $= 9.1 \times 10^{-31}$ kg, charge of electron $= -1.6 \times 10^{-19}$ C)

6 Capacitors and d.c.

6.1 Introduction

Capacitors are used for storing charge. They vary in shape, size and type according to their applications, Figure 6.1.

6.2 Capacitors

The circuit in Figure 6.2 can be used to illustrate some properties of a capacitor in a d.c. circuit.

When switch S is turned to position 1, the lamp L_1 glows brightly, dims and then goes out. The current is at a maximum immediately after the switch is closed and rapidly falls to zero, Figure 6.3(a). If the switch S is then turned to position 2, the lamp L_2 glows brightly, dims and then goes out even though the battery is no longer part of the circuit. The current is once again at a maximum immediately after the switch is closed and rapidly falls to zero although in this case the current is in the **opposite direction**, Figure 6.3(b).

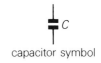

Figure 6.1 Capacitors

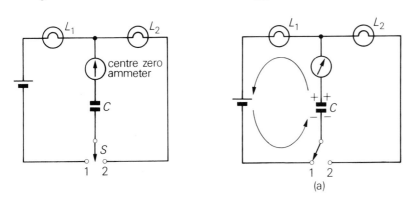

Figure 6.2

Figure 6.3

When a current passes for a certain time, a quantity of charge is transferred from one part of the circuit to another. Figure 6.3(a) shows the battery transferring electrons from the top plate to the lower plate of the capacitor. We say that the capacitor has been **charged**. In Figure 6.3(b) the capacitor releases the stored charge; electrons flow back to the top plate and we say that the capacitor is **discharging**.

A capacitor consists of two metal plates (or foils), separated by an insulating material. Figure 6.4 shows the construction of a typical capacitor.

Charging the capacitor required a potential difference, so work was done in transferring charge between the plates. The relationship between the quantity of charge Q transferred to the plates and the potential difference V between them is investigated using the apparatus shown in Figure 6.5. The parallel plate capacitor shown consists of two metal foil plates whose separation and area of overlap can be varied. One form of this apparatus is called the 'Aepinus Capacitor', as shown; other forms have horizontally placed sheets of foil with insulating separators.

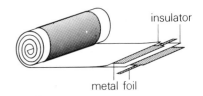

Figure 6.4 Capacitor construction

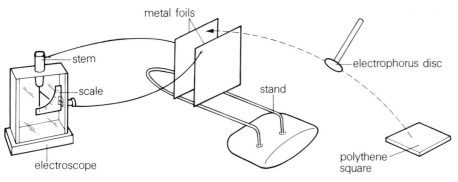

Figure 6.5

As charge is supplied to the capacitor plate, a potential difference is set up between the plates. Charge can be supplied using an electrophorus. When the electrophorus disc is placed on the negatively charged polythene square and earthed, the disc becomes positively charged by induction. An equal quantity of charge is transferred by the disc to the capacitor plate each time.

As more and more charge is transferred to the capacitor, more and more work is being done by the experimenter as he/she charges the capacitor. As the quantity of charge stored in the capacitor builds up, so too does the electric field between the parallel plates.

If one plate of the capacitor is connected to the case and the other to the stem of the electroscope, the deflection of the gold leaf gives a measure of the potential difference between the plates.

As more charge is transferred to the capacitor, the potential difference increases and is measured by the number of divisions of deflection of the gold leaf. This experiment provides results of the form shown in Table 1.

Quantity of charge transferred (number of charge transfers)	Q	1	2	3
potential difference (electroscope divisions)	V	2	4	6

Table 1

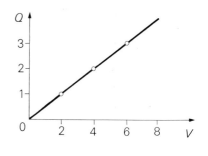

Figure 6.6

A graph of Q versus V has a constant slope and passes through the origin (Figure 6.6), so that Q and V are related by the equation, $Q = \text{constant} \times V$

This can be written

$$\frac{Q}{V} = \text{constant} \quad \Rightarrow \quad \frac{Q}{V} = C$$

The constant C defines the capacitance of the capacitor

$$\text{capacitance } (C) = \frac{\text{charge } (Q)}{\text{potential difference } (V)}$$

When Q is measured in coulombs and V is measured in volts, the capacitance is measured in farads.

If a charge of 1 coulomb (1 C) is transferred to the plate of a capacitor and this results in a potential difference between the plates of 1 volt (1 V), then the capacitance is 1 farad (1 F).

In practice one farad is a very large capacitance and most capacitors have much smaller values which may be expressed in microfarads (μF), nanofarads (nF) or picofarads (pF).

$$1\mu F = 1 \times 10^{-6} F \qquad\qquad 1nF = 1 \times 10^{-9} F \qquad\qquad 1pF = 1 \times 10^{-12} F$$

Example 1

What quantity of charge is needed to charge a 2.0 microfarad capacitor to a potential difference of 12 volts?

$$C = \frac{Q}{V} \quad \text{where } C = 2.0 \times 10^{-6}\,\text{F} \quad V = 12\,\text{V}$$

$$\Rightarrow \quad 2.0 \times 10^{-6} = \frac{Q}{12}$$

$$\Rightarrow \quad 12 \times 2.0 \times 10^{-6} = Q$$

$$\Rightarrow \quad Q = 24 \times 10^{-6}$$

24μ C of charge is required to charge the capacitor

Example 2

A charge of 3.0×10^{-12} C transferred to the plate of a capacitor produces a potential difference of 2.0 mV. What is the capacitance of the capacitor?

$$C = \frac{Q}{V} \quad \text{where } Q = 3.0 \times 10^{-12}\,\text{C} \quad V = 2.0 \times 10^{-3}\,\text{V}$$

$$\Rightarrow C = \frac{3.0 \times 10^{-12}}{2.0 \times 10^{-3}}$$

$$\Rightarrow C = 1.5 \times 10^{-9}$$

The capacitor has a capacitance 1.5 nF

6.3 Factors affecting capacitance

Area of plates

Using the capacitor system, the area of overlap of the plates can be varied in order to find out how the capacitance is affected, Figure 6.7.

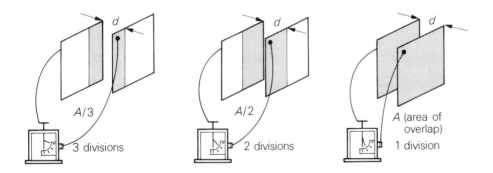

Figure 6.7 Area of capacitor plates

The distance d between the plates is kept constant throughout the experiment. The quantity of charge Q is kept constant by transferring a fixed amount at the beginning of the experiment. The area of overlap can be altered by sliding one plate relative to the other. As the area of overlap of the metal plates increases, the electroscope deflection decreases. Since the deflection indicates the potential difference V between the plates and $C = Q/V$, then as V decreases, the capacitance C increases. This means that the capacitance depends on the area of overlap. The experiment reveals the following pattern of results, Figure 6.8.

overlap area A	increases
p.d. (divisions) V	decreases
charge (Q)	remains constant
capacitance (Q/V)	increases

Figure 6.8

In fact the capacitance of a capacitor is directly proportional to the area of overlap of its plates

$C \propto A$

The dependence of capacitance on the area of overlap has been used in the variable capacitor used for tuning a radio, Figure 6.9. Movement of the semicircular metal plates (vanes) relative to each other produces a variation of the capacitance.

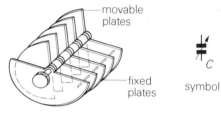

Figure 6.9

Distance between capacitor plates

Using the same apparatus, keeping the area of overlap A constant and the quantity of charge Q on the system constant, the distance d between the plates can be varied, Figure 6.10.

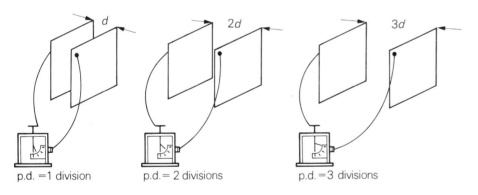

p.d. =1 division   p.d.= 2 divisions   p.d.=3 divisions

Figure 6.10 Distance between capacitor plates

As the distance between the plates is varied, the potential difference between the plates is noted. Typical results are show in Figure 6.11.

plate separation d	increases
p.d. (divisions) (V)	increases
charge (Q)	remains constant
capacitance (Q/V)	decreases

Figure 6.11

As the plate separation is increased, the potential difference V increases; since $C = Q/V$, this means that the capacitance decreases. In fact the capacitance varies inversely as the plate separation d,

$$C \propto \frac{1}{d}$$

This dependence of capacitance on plate separation is used in the 'trimmer' capacitor (used to make very fine adjustments to the tuning of a radio), Figure 6.12. The spacing of the metal plates is adjusted by turning the screw and hence the capacitance varies.

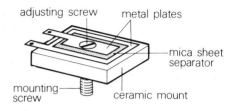

Figure 6.12 Trimmer capacitor

If these two results are combined, the capacitance of two parallel metal plates in air is described by the equation.

$$C \propto \frac{A}{d} \quad \Rightarrow \quad C = k \frac{A}{d} \qquad$$
where A = area of plate overlap
d = separation of plates
k = a constant

Material between capacitor plates

We can observe the effect on the capacitance of inserting different insulating materials between the plates of the capacitor as in Figure 6.13.

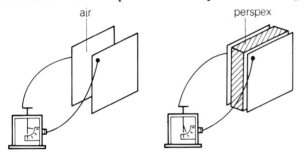

Figure 6.13 Material between capacitor plates

With air between the plates, the capacitor is charged to a particular potential difference using a fixed quantity of charge. The separation of the plates and the area of overlap are kept constant. When a piece of perspex is inserted in the space between the plates, the electroscope deflection (which is a measure of the potential difference V), decreases. Since the capacitance C is given by $C = Q/V$ and Q is kept constant then, if V decreases, C must increase.

The dependence of capacitance on the type of material between the plates of the capacitor is expressed by the constant k in the equation

$$C = k \frac{A}{d}$$

The insulating material between the plates is called a dielectric and the constant k is the factor by which the capacitance is increased when that material is inserted between the plates. The k values given in Table 1 show that different materials influence the capacitance by different amounts.

material between plates	k factor
air	1.0
waxed paper	2.7
polyester	3.8
mica	7.0

Table 1

For general purposes, capacitors use waxed paper as the dielectric between the two metal foils as already shown in Figure 6.4. Other materials are chosen for special applications.

Another type of capacitor, the electrolytic capacitor has plates of aluminium foil, one of which has an oxide coating. There is a layer of chemical-soaked paper (electrolyte) between the foils. This type of capacitor is well sealed to prevent leakage, Figure 6.14. It is important to connect this type of capacitor with the correct polarity in a d.c. circuit as it can be damaged by incorrect connection. The electrolytic capacitor has its own symbol.

Capacitors have a 'working voltage' printed on them. This is the maximum recommended p.d. which should be applied to them without the risk of the insulation of the dielectric breaking down. This could result in a large current surge between the plates and rapid heating, leading to a dangerous explosion.

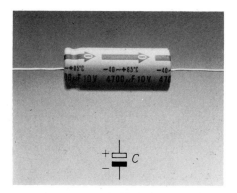

Figure 6.14 Electrolytic capacitor

6.4 Energy stored in a capacitor

In section 6.2, it was shown how the p.d. between the plates of a capacitor increased as the quantity of charge stored increased, Figure 6.15. The quantity of work done (or energy stored) can be found from the area under the Q versus V graph

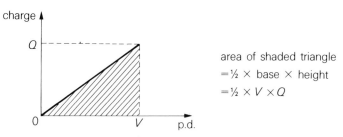

area of shaded triangle
$= \frac{1}{2} \times$ base $\times$ height
$= \frac{1}{2} \times V \times Q$

Figure 6.15 Energy stored in a capacitor

The energy stored in a capacitor $= \frac{1}{2} Q V$ joules.

Alternatively, since $C = \dfrac{Q}{V}$

$$\text{energy} = \tfrac{1}{2} C V^2 = \tfrac{1}{2} \frac{Q^2}{C}$$

Example 3

A $100\,\mu$F capacitor is connected to a 12 V supply. Calculate the charge on the capacitor and the energy stored.

$$C = \frac{Q}{V} \quad \text{where } C = 100\,\mu\text{F and } V = 12\,\text{V}$$

$$\Rightarrow \quad Q = C V = 100 \times 10^{-6} \times 12$$
$$\Rightarrow \quad Q = 1.2 \times 10^{-3} \text{ coulombs}$$

$$\text{energy} = \tfrac{1}{2} Q V$$
$$= \tfrac{1}{2} \times 1.2 \times 10^{-3} \times 12 = 7.2 \times 10^{-3} \text{ joules}$$

The charge on the capacitor is 1.2×10^{-3} C and the energy stored is 7.2 mJ

6.5 Capacitors in d.c. circuits

Charging a capacitor

The circuit in Figure 6.16 can help us to 'see' more clearly the current variation which takes place when a capacitor is charged from a d.c. supply.

The capacitor C is initially uncharged. When switch S is closed, electrons flow in an anti-clockwise direction: the lower plate becomes negatively charged and the upper plate positively charged. Thus there is a growth of potential difference across the capacitor. This potential difference opposes further flow of charge, thus reducing the charging current, Figure 6.17.

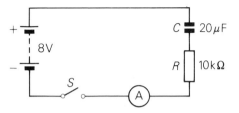

Figure 6.16 Charging a capacitor

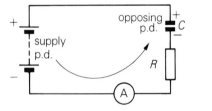

Figure 6.17

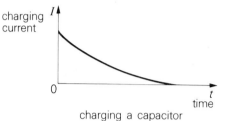

Figure 6.18 Charging a capacitor

Eventually enough charge has flowed to C so that the potential difference across C is equal to the supply potential difference (8 volts, in this case). At this stage, the potential difference across the capacitor is equal and opposite to that of the supply and so the net charge flow is zero. The variation of charging current with time is shown in Figure 6.18.

Discharging a capacitor

The discharge of a capacitor can be observed in detail if it is first fully charged to a given potential difference with switch S at position 1 in the circuit in Figure 6.19.

The voltmeter indicates when the capacitor has been charged to the required potential difference. Switch S is then moved to position 2. Since the capacitor has a large potential difference across it, the charge flow from it is a maximum at the start. As charge flows from one capacitor plate to the other through R and the ammeter (Figure 6.20(a)), the potential difference across them reduces and hence the discharge current reduces. The variation of discharge current with time is shown in Figure 6.20(b).

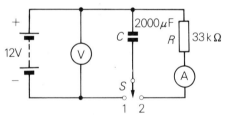

Figure 6.19

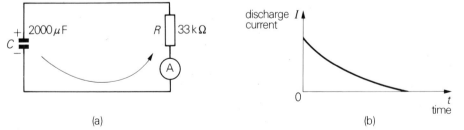

(a)   (b)

Figure 6.20

When a larger resistor R is substituted in the circuit, the current is reduced. Therefore the rate of flow of charge from the capacitor plates is reduced, with the result that it takes longer for the capacitor to discharge. Similarly, if a larger capacitor C is substituted then the total charge Q that it can store is increased; it therefore takes longer for the capacitor to discharge, Figure 6.21.

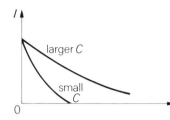
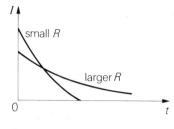

Figure 6.21 Capacitor discharge

These graphs of discharge show the variation of current in the opposite direction. If we show the charge and discharge currents for a capacitor using the same set of axes, we obtain the curves shown in Figure 6.22.

Figure 6.22 Charging and discharging a capacitor

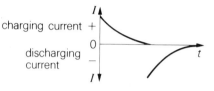

Charging and discharging capacitors – equations

As the charge builds up on the plates of a capacitor, the potential difference between them increases until it reaches a maximum value V_c Figure 6.23.

The charging current I is the rate at which the capacitor gains charge and can be written $\frac{dQ}{dt}$.

$$I = \frac{dQ}{dt}$$

But $Q = C \times V$ Where Q is the charge

$\Rightarrow$ $I = \frac{d(CV)}{dt}$ C is the capacitance

     V is the p.d. across the capacitor

$\Rightarrow$ $I = C\frac{dV}{dt}$ (since C is constant)

Figure 6.23

The current depends on the rate at which the potential difference across the capacitor changes. Comparing this equation with the information given in Figure 6.23, we can see that when the time t is large, then the slope of the voltage-time graph becomes zero.

i.e. $\frac{dV}{dt} = 0 \Rightarrow I = 0$ at this stage

The capacitor-charging current reduces to zero as the potential difference across the plates reaches its maximum value.
Consider now the circuit shown in Figure 6.24.

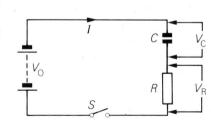

Figure 6.24

At any instant the current is I. Now the source V_0 drives the current in a clockwise direction and C and R both oppose the current, so that V_C and V_R must both oppose V_0.

$\Rightarrow V_0 = -(V_C + V_R)$ where V_C is the p.d. across the capacitor

     V_R is the p.d. across the resistor

By using the equation $I = C\,dV/dt$, it is possible to describe how the potential difference across the components in this circuit varies with time.

For the capacitor, $V_C = \dfrac{Q}{C}$

For the resistor, $V_R = I\,R$

a) At the start there is no charge on the capacitor, so Q is zero.

$$\therefore\; V_C = \frac{0}{C} = 0$$

$$\Rightarrow\; V_0 = -(0 + V_R)$$

$$\Rightarrow\; V_R = -V_0$$

$$I = -\frac{V_0}{R} \text{ (the maximum value)}$$

b) When V has stopped charging and the current I is zero,

$$\frac{dV}{dt} = 0$$

$$I = 0$$

$$V_R = I\,R$$

$$\Rightarrow\; V_R = 0$$

$$\Rightarrow\; V_0 = -(V_C + 0)$$

$$\Rightarrow\; V_C = -V_0$$

Figure 6.25 illustrates these variations in potential difference.

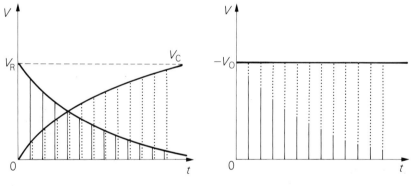

Figure 6.25

This is summarised in Table 2.

	at start	when V no longer changes
V_R	$-V_0$	0
V_C	0	$-V_0$
I	$-\dfrac{V_0}{R}$	0

Table 2

Example 4

a) Calculate the maximum charging current in the circuit shown.
b) What are the final potential differences across the capacitor and across the resistor?

a) $V_0 = -(V_C + V_R)$ where $V_C = \dfrac{Q}{C}$
and $V_R = I \times R$

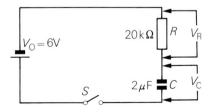

The current has its maximum value at the start. At this time, there is no charge on the capacitor, $Q = 0 \Rightarrow V_C = 0$

$$\Rightarrow V_0 = -(0 + V_R)$$
$$\Rightarrow V_0 = I_{max} \times R \quad \text{where } R = 20 \times 10^3 \, \Omega$$
$$V_0 = 6 \, V$$

$$\Rightarrow I_{max} = \frac{V_0}{R}$$
$$= \frac{6}{20 \times 10^3} = 3 \times 10^{-4} = 300 \times 10^{-6}$$

The maximum current is $300 \, \mu A$

b) The charge flows until the capacitor is fully charged, and then the current I is zero.

$$V_R = I \times R$$
$$\text{When } I = 0, \; V_R = 0$$
$$V_0 = -(V_C + V_R)$$
$$= -V_C$$

The final p.d. across C is $V_C = 6$ volts.
The final p.d. across R is $V_R = 0$ volts.

6.6   Combining capacitors

Capacitors in parallel

If two capacitors, C_1 and C_2 are connected in parallel as shown in Figure 6.26, they both become charged to the same potential difference as the supply voltage, V_0. Using the equation for capacitance $Q = CV$

$$\Rightarrow Q_1 = C_1 \times V_0$$
$$\Rightarrow Q_2 = C_2 \times V_0$$

We can derive the value of the equivalent parallel capacitor, C_p, by considering the system replaced by the following circuit, Figure 6.27.

$$Q = C_p \times V_0$$

This single capacitor will hold the same amount of charge as C_1 and C_2

$$Q = Q_1 + Q_2$$
$$\Rightarrow C_p V_0 = C_1 V_0 + C_2 V_0$$
$$C_p = C_1 + C_2$$

By placing capacitors in parallel we can increase the total capacitance in a circuit. As Figure 6.28 shows, we are effectively increasing the area of capacitor plates and thus the size of the capacitor.

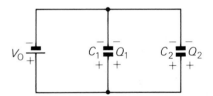

Figure 6.26

Figure 6.27

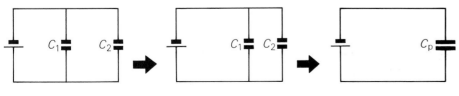

Figure 6.28

Example 5

Calculate the equivalent capacitance when a $20\,\mu\text{F}$ capacitor and a $100\,\mu\text{F}$ capacitor are connected in parallel.

$$C_p = C_1 + C_2 \quad \text{where } C_1 = 20\,\mu\text{F}; C_2 = 100\,\mu\text{F}$$
$$\Rightarrow C_p = 20 \times 10^{-6} + 100 \times 10^{-6}$$
$$= 120 \times 10^{-6}$$

The equivalent capacitance is $120\,\mu\text{F}$

Example 6

What capacitor in parallel with a $25\,\mu\text{F}$ capacitor will give a total capacitance of $100\,\mu\text{F}$?

$$C_p = C_1 + C_2 \quad \text{where } C_1 = 25\,\mu\text{F}; C_2 = ?; C_p = 100\,\mu\text{F}$$
$$\Rightarrow 100 \times 10^{-6} = 25 \times 10^{-6} + C_2$$
$$\Rightarrow C_2 = 75 \times 10^{-6}$$

A capacitor of $75\,\mu\text{F}$ should be placed in parallel

Capacitors in series

The two capacitors, C_1 and C_2 in Figure 6.29 are connected in series. If we consider the potential difference across each capacitor, we can write
$V_0 = V_1 + V_2$

C_s is the equivalent series capacitor.

Because $V = \dfrac{Q}{C}$ we can write $\dfrac{Q}{C_s} = \dfrac{Q}{C_1} + \dfrac{Q}{C_2}$

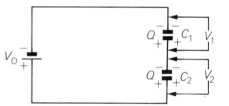

Figure 6.29

The charge on both capacitors is the same since charging the capacitors involves the battery moving charge Q from the bottom plate of C_2 and adding charge Q to the top plate of C_1. The top plate of C_2 has a quantity of negative charge Q and the bottom plate of C_1 has a quantity of positive charge Q.

$$\Rightarrow \frac{1}{C_s} = \frac{1}{C_1} + \frac{1}{C_2}$$

The placing of the two capacitors in series has the same effect as increasing the distance between the plates. The charges on the two middle plates of the combination are equal and opposite and so are neutralized. This means that these two middle plates are not really involved, and the plates of the equivalent capacitor have a greater separation.

Since $C \propto 1/d$, the equivalent capacitance of two capacitors in series must be less than either individually.

Example 7

What is the effective capacitance when a $20\,\mu\text{F}$ capacitor is placed in series with a $30\,\mu\text{F}$ capacitor?

$$\frac{1}{C_s} = \frac{1}{C_1} + \frac{1}{C_2} \quad \text{where } C_1 = 20\,\mu\text{F}; C_2 = 30\,\mu\text{F}$$
$$\Rightarrow \frac{1}{C_s} = \frac{1}{20 \times 10^{-6}} + \frac{1}{30 \times 10^{-6}} = \frac{3+2}{60 \times 10^{-6}} = \frac{5}{60 \times 10^{-6}}$$
$$\Rightarrow C_s = \frac{60 \times 10^{-6}}{5} = 12 \times 10^{-6}$$

The effective capacitance is $12\,\mu\text{F}$

Example 8

What capacitor is required in series with a $10\,\mu F$ capacitor in order to provide an equivalent capacitor of $6\,\mu F$?

$$\frac{1}{C_s} = \frac{1}{C_1} + \frac{1}{C_2} \quad \text{where } C_1 = 10\,\mu F;\ C_2 = ?;\ C_s = 6\,\mu F$$

$$\Rightarrow \frac{1}{6 \times 10^{-6}} = \frac{1}{10 \times 10^{-6}} + \frac{1}{C_2}$$

$$\Rightarrow \frac{1}{6 \times 10^{-6}} - \frac{1}{10 \times 10^{-6}} = \frac{1}{C_2}$$

$$\Rightarrow \frac{1}{C_2} = \frac{10 - 6}{60 \times 10^{-6}} = \frac{4}{60 \times 10^{-6}}$$

$$\Rightarrow C_2 = \frac{60 \times 10^{-6}}{4} = 15 \times 10^{-6}$$

A capacitor of value $15\,\mu F$ is required in series

Summary

The potential difference across two parallel conducting plates is directly proportional to the charge on them.

Capacitance is defined by the ratio charge to potential difference.

$$C = \frac{Q}{V}$$

1 farad = 1 coulomb per volt; $1\,\text{F} = 1\,\text{C V}^{-1}$

The capacitance of a parallel plate capacitor is directly proportional to the area of overlap of the plates and inversely proportional to the distance between the plates; it depends also on the dielectric material between the plates.

The energy stored in a capacitor of capacitance C farads, charged to a p.d. of V volts is given by $E = \frac{1}{2}C V^2$ joules.

When a capacitor is charged from a d.c. supply, the current is at first large and gradually reduces to zero; the p.d. across the capacitor rises from zero to a maximum value during this time.

For capacitors in parallel in a circuit, the total capacitance may be calculated from:

$$C_p = C_1 + C_2 + C_3 + \dots$$

For capacitors in series in a circuit, the total capacitance may be calculated from:

$$1/C_s = 1/C_1 + 1/C_2 + 1/C_3 + \dots$$

Problems

1   What is the relationship between the charge Q on a capacitor and the p.d. V across its plates? Describe a simple experiment to find this relationship.

2   Define capacitance in terms of charge and p.d. Write an equation which includes units as well as symbols.

3   Describe how to find experimentally the effect of the following factors on the capacitance of a parallel plate capacitor.
   a) area of plates,
   b) separation of plates,
   c) dielectric between plates.

4   State the relationship between capacitance C of a parallel plate capacitor and the area of overlap A and separation d of its plates.

5   What quantity of charge is needed to charge a 5 μF capacitor to a p.d. of 12 V?

6   A charge of 5×10^{-12} C transferred to the plates of a capacitor produces a p.d. of 10 mV. What is the capacitance of the capacitor?

7   A parallel plate capacitor is connected to a battery. What happens to its charge and the p.d. across its plates when a slab of dielectric is inserted between its plates?

8   A smoothing capacitor in a low voltage power supply has to store 6 mC of charge when the p.d. across it is 12 V. What capacitance should it have?

9   Draw a circuit which will demonstrate the charge and discharge of a capacitor. Draw a current-time graph for charge and discharge of a capacitor in a d.c. circuit.

10   Draw the p.d.-time graphs for charge and discharge of a capacitor as described in Question 9.

11   Draw current-time graphs for the charging of a large and a small value capacitor for a constant voltage d.c. supply.

12   Draw current-time graphs to illustrate the effect of a large and a small resistor on the charging of a capacitor from a constant voltage d.c. supply.

13   Consider the circuit shown

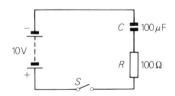

   a) What is the maximum charging current?
   b) What is the final p.d. across C?
   c) How much charge is transferred to the capacitor?

14   A 2000 μF capacitor is connected to a 15 V d.c. supply. Calculate the charge and energy stored in the capacitor.

15 a) Write the expressions for the combination of two capacitors in series and in parallel.
b) Calculate the total capacitance of the combination shown.

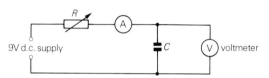

16 The capacitor C is charged with a steady current of 1 mA by carefully adjusting the variable resistor R.

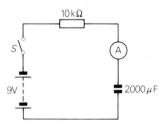

The voltmeter reading is taken every ten seconds. The results are shown in the table.

Time (s)	0	10	20	30	40
Voltmeter reading (V)	0	1.9	4.0	6.2	8.1

Plot a graph of charge against voltage for the capacitor and hence find its capacitance.

SCEEB

17 To study the charging of a capacitor the circuit shown is used.

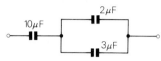

a) Describe the response of the ammeter after switch S is closed.
b) How would you know when the potential difference across the capacitor is at its maximum?
c) Suggest a suitable range for the ammeter.
d) If the 10 kΩ resistor is replaced by one of larger resistance, what will be the effect on the maximum potential difference across the capacitor?

SCEEB

18 In the circuit below, the neon lamp flashes at regular intervals.

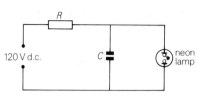

The neon lamp requires a potential difference of 100 V across it before it will conduct and flash. It continues to glow until the potential difference across it drops to 80 V. While it is lit, its resistance is very small compared with R.
a) Explain why the neon lamp flashes regularly.
b) Suggest **two** methods of decreasing the flash rate.

SCEEB

19 The charging of a capacitor is studied using the circuit shown. The ammeter A is a centre-zero instrument and a constant d.c. supply of 6 volts is used.

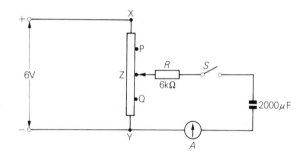

a) The sliding contact Z is set at the mid-point of XY and switch S is closed.
　i) Calculate the final charge on the capacitor C.
　ii) Sketch a graph showing how the charging current varies with time.
b) With the capacitor fully charged as in (**a**) describe how the current through the milliameter changes when the sliding contact Z is moved, in turn, to a new position
　i) P (nearer to X)
　ii) Q (nearer to Y).
c) At the end of the experiment switch S is opened and a short conducting wire is connected directly across the plates of the capacitor.
Describe what happens to the energy that was stored in the capacitor.

SCEEB

7 Inductors and d.c.

7.1 Electromagnetic induction

When a current passes through a coil of wire, a magnetic field is formed around the coil. The pattern of the magnetic field is similar to that around a bar magnet, Figure 7.1. The magnetic field is associated with the current in the wire. We now look at the current that can be formed by a magnetic field.

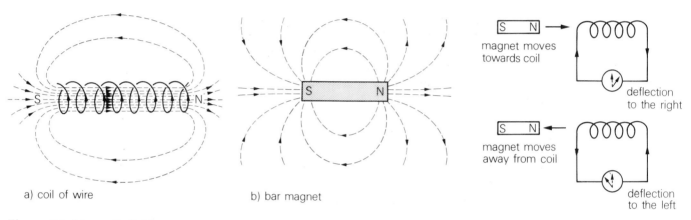

a) coil of wire   b) bar magnet

Figure 7.1 Magnetic fields

Figure 7.2 Magnetic induction in a coil

A current can be **induced** in a coil of wire that is connected only to a centre-zero milliammeter when a magnet is moved towards or away from the coil, Figure 7.2. This is an example of **electromagnetic induction.**

The following points can be observed from the experiment.

1 A current passes **only** when the magnet is moving, i.e. when the magnetic field around the coil is changing.

2 Movement of the magnet towards the coil produces a current in the reverse direction from that resulting from movement of the magnet away from the coil.

3 The direction of the current in the coil is also reversed by reversing the poles of the magnet.

The temporary current induced by the movement of the magnet results from an electromotive force in the coil which is called an **induced e.m.f.**

The following factors affect the magnitude of the induced e.m.f.

1 The length of the conductor
If a coil having a greater number of turns is used, the meter deflection is greater, indicating a greater induced e.m.f.

2 The strength of the magnetic field
A stronger magnetic field results in a greater induced e.m.f.

3 The speed of movement
The speed of movement governs the rate at which the magnetic field changes. The faster the magnetic field changes, the greater is the induced e.m.f.

The direction of the induced e.m.f. is shown for various movements of the magnet in Figure 7.3.

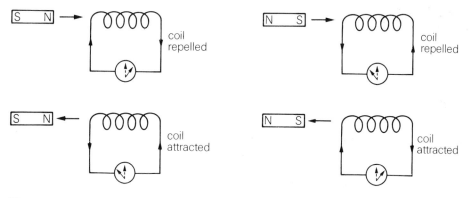

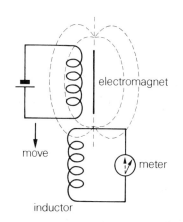

Figure 7.3 Direction of induced e.m.f.

The diagram shows that there is a force on the coil which is pushing it away from the magnet when the magnet approaches it. Conversely, there is a force on the coil which is pulling it towards the magnet when the magnet is moved away from it. The force is an electromotive force which induces a polarity on the coil to resist the motion of the magnet, and which causes the small current to flow briefly through a meter connected to the coil.

These results are summarised in **Lenz's Law:**

'The direction of the induced e.m.f. is such that it opposes the change producing it'.

The e.m.f. is induced simply by the change in the magnetic field around the coil; the coil is normally called an **inductor**.

The magnetic field can be changed by replacing the magnet by an electromagnet and moving that instead, Figure 7.4.

The magnetic field can also be changed merely by turning the current on and off in the electromagnet if it is placed near the inductor: this is called **mutual induction**. Figure 7.5 shows what happens when switch *S* in the electromagnet circuit is switched on and off.

Figure 7.4 Induction using a moving electromagnet

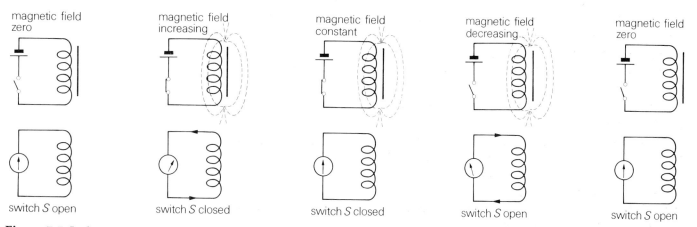

Figure 7.5 Induction using a stationary electromagnet

Notice that when the field of the electromagnet is increasing, the induced current is in one direction; when the field of the electromagnet is decreasing, the induced current is in the opposite direction.

The induced e.m.f. is greater if an iron core is inserted in the coil. The iron core increases the strength of the magnetic field through the coils when the current is switched on.

7.2   Inductors

When a current is passed through a coil of wire, any change in the current sets up a magnetic field which induces an e.m.f. in the coil itself. This induced e.m.f. opposes the change, and is called **self-induction**. Thus any coil can act as an inductor when the current through it is changed. The symbols for an air-core inductor and an iron-core inductor are shown on the right.

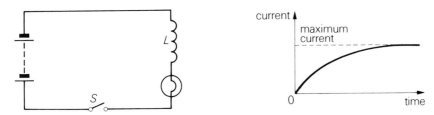

Figure 7.6 Inductor circuit

Figure 7.7 Growth of current in an inductor

In Figure 7.6, a lamp is connected to an inductor L and a battery through a switch S. If the lamp is connected only to the battery and switched on, it becomes fully lit immediately. The lamp in the diagram takes a noticeable time to light fully when the switch is closed because the inductor L slows the build-up of current. The increase in the current in the inductor produces an increasing magnetic field which results in a **self-induced** e.m.f. in the inductor.

From Lenz's Law, the self-induced e.m.f. **opposes** the change producing it. So the induced e.m.f. opposes the supply e.m.f. and the rate of increase in the current is slowed, Figure 7.7.

After a short time the current reaches a steady, maximum value. The current is then no longer changing, so there is no changing magnetic field and no self-induced e.m.f. to oppose the supply e.m.f.

When switch S is opened, the current falls very rapidly to zero; this causes a rapid reduction in magnetic field and a large self-induced e.m.f., the size of which can be demonstrated by means of a neon lamp, Figure 7.8.

A neon lamp contains neon gas at low pressure which only conducts electricity when the p.d. applied is sufficiently high.

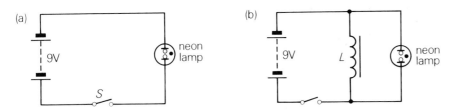

Figure 7.8

In Figure 7.8(a) when switch S is closed, the neon lamp does not light: the 9 V supply is insufficiently high.

When switch S is closed in Figure 7.8(b) there is also no observable effect as the current builds up to its maximum value. But when the switch in this circuit is opened, the neon lamp is seen to flash briefly. The self-induced e.m.f. from the inductor L is sufficient to cause the neon to conduct electricity. The extra energy causes the neon gas to emit a red light.

The work done in building up the current in the inductor was stored in the magnetic field of the inductor. When the switch S was opened, the energy stored in this magnetic field was released to provide the e.m.f. that lit the neon lamp.

7.3   Factors affecting inductance

Core material

Figure 7.9 shows the construction of a typical inductor. A length of insulated wire is wrapped in many turns around an insulating former. The former is hollow if the inductor is air-cored, Figure 7.9(a), but a pair of C-shaped pieces of soft iron may be inserted as in Figure 7.9(b) to make it iron-cored.

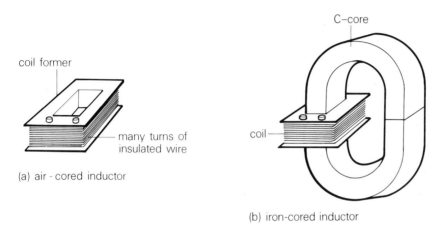

(a) air - cored inductor

(b) iron-cored inductor

Figure 7.9 Inductors

The circuit in Figure 7.10 may be used to demonstrate the effect of the core on the inductor. L_1 is air-cored and L_2 is identical but has a pair of C-cores forming an iron core. The two lamps B_1 and B_2 are identical. When switch S is closed, B_1 is fully lit before B_2 because the iron-cored inductor L_2 causes a slower growth of current. We say that the inductance of L_2 is greater: the soft iron core increases the strength of the magnetic field and hence of the self-induced e.m.f.

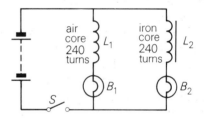

Figure 7.10 Effect of inductor core

Eddy currents

The changing magnetic field of the inductor coil results in an induced e.m.f. in the soft iron core which is a conductor of electricity. So currents are produced in the core called **eddy currents**, Figure 7.11. These eddy currents produce heat in the core material which results in energy loss. The loss can be reduced by using a laminated C-core in which a number of thin sheets of iron are glued together with thin layers of insulation between them, Figure 7.12.

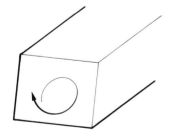

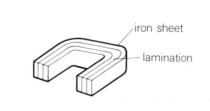

Figure 7.11 Eddy currents in core   **Figure 7.12** Laminated C-core

Since no current passes through the layers of insulation, eddy currents are reduced.

The energy lost in magnetizing and demagnetizing the iron core is also reduced by using soft iron for the core. Soft iron does not retain its magnetism and needs little energy to be magnetized and demagnetized.

Number of turns in inductor coil

The circuit in Figure 7.13 is used to compare the time taken for the two bulbs to light up when the number of turns of wire wound on the inductor L is varied. About 8 metres of insulated wire are connected between the points X and Y on the circuit. As this has an appreciable resistance, the variable resistor R is first adjusted so that B_1 and B_2 are equally bright when S is closed and the current is steady. The difference in time between B_1 and B_2 lighting after S is closed is then measured for every extra twenty turns of wire wound on the inductor L. The experiment shows that the delay time for bulb B_1 to light increases as more turns of wire are wound on to the inductor former.

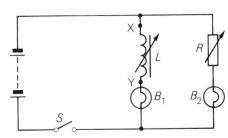

Figure 7.13

Dimensions of inductor coil

The value of the inductance of a coil is influenced by many factors related to its shape. For example, the uncoiled wire of the inductor in Figure 7.13 had no appreciable inductance; when it was wound on the former, the value of the inductance had completely changed.

Factors which can influence the inductance value are as follows:
a) the spacing between the turns
b) the wire size (thickness)
c) the number of layers of windings
d) the diameter of the coil
e) the construction of the winding

An inductor coil

7.4 Inductors – some equations

From a given coil, the value of the induced e.m.f. depends only on the rate at which the magnetic field changes, i.e. on the rate of change of the current through the coil.

The induced e.m.f. E is proportional to the rate of change of current dI/dt.

$$E \propto -\frac{dI}{dt}$$

Because the self-induced e.m.f. opposes the change of current, the equation includes a minus sign.

$$\Rightarrow E = -k\frac{dI}{dt}$$

where the constant k depends on the properties of the inductor coil: it is called the inductance L.

$$\Rightarrow E = -L\frac{dI}{dt}$$

$$\Rightarrow L = -\frac{E \text{ (volts)}}{dI/dt \text{ (amps per second)}}$$

The unit of inductance is the henry (H) where

$$1\,\text{H} = 1\,\text{V}\,\text{s}\,\text{A}^{-1}$$

If the induced e.m.f. is 1 volt when the current changes at a rate of 1 ampere per second, the inductance is 1 henry.

Figure 7.14 Michael Faraday

Consider again the current-time graph for the growth of current in a circuit containing an inductor, Figure 7.15.

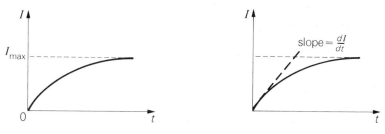

Figure 7.15

Figure 7.16

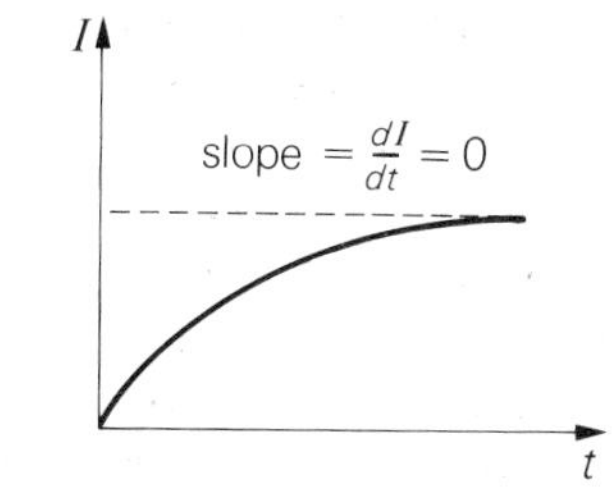

Figure 7.17

a) The current is zero at the start when $t = 0$; but at the start, the current is changing at the greatest rate because the slope dI/dt of the graph is at a maximum, Figure 7.16.

Because $E = -L dI/dt$, the value of the self-induced e.m.f. E opposing the current is also at a maximum.

b) After a certain time, the current reaches a maximum value I_{max}; the slope of the graph dI/dt has then become zero, Figure 7.17.

Because $E = -L dI/dt$, the value of the self-induced e.m.f. E is also zero, there is no opposition to the current, and steady conditions exist.

These two conditions are used in calculating the inductance of a circuit connected to a d.c. supply.

Example 1

Figure 7.18 shows an inductor of 3 H connected to a d.c. supply. When switch S is closed, a self-induced e.m.f. of 9 V is produced. Find the maximum rate at which the current increases.

$$E = -L\frac{dI}{dt} \quad \text{where} \quad L = 3\,\text{H}; \quad E = 9\,\text{V}$$

The e.m.f. E is negative because it opposes the supply p.d.

$$\Rightarrow \quad -9 = -3\frac{dI}{dt}$$

$$\Rightarrow \quad \frac{dI}{dt} = 3$$

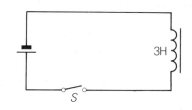

Figure 7.18

The current increases at a maximum rate of 3 A s⁻¹

Example 2

When the circuit in Figure 7.19 is switched off, the current decreases at a maximum rate of $1.0 \times 10^4\,\text{A s}^{-1}$. Calculate the maximum self-induced e.m.f.

$$E = -L\frac{dI}{dt} \quad \text{where} \quad L = 4.0 \times 10^{-3}\,\text{H}$$

$$\frac{dI}{dt} = -1.0 \times 10^4\,\text{A s}^{-1}$$

dI/dt is negative because the current is decreasing.

$$\Rightarrow \quad E = -4.0 \times 10^{-3} \times 1.0 \times 10^4 = 40$$

The maximum self-induced e.m.f. is 40 V

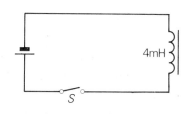

Figure 7.19

7.5 Inductors in d.c. circuits

A practical inductor has a certain amount of resistance due to the windings of its coil. It is convenient to show an inductor in a circuit diagram as having both inductance and resistance, Figure 7.20.

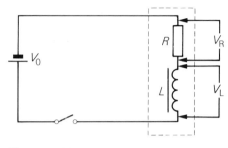

Figure 7.20

The self-induced e.m.f. V_L depends on the rate at which the current changes.

$$V_L = -L\frac{dI}{dt}$$

This equation is similar to $I = CdV/dT$ which was used on page 69 for a capacitor but it involves a minus sign because the e.m.f. opposes the change of current.

At any instant the current is I; this is driven by the p.d. V_0 of the source in a clockwise direction round the circuit.

The p.d.'s V_L and V_R both oppose the current, so they must both oppose V_0:

$$V_0 = -(V_L + V_R)$$

also $V_R = -IR$ and $V_L = -L\frac{dI}{dt}$

a) From Figure 7.21, at the start $I = 0$

$\Rightarrow \qquad V_R = -IR = 0$

$\Rightarrow \qquad V_0 = -(0 + V_L)$

$\Rightarrow \qquad V_L = -V_0$

The p.d. V_R across the resistor is zero.
The p.d. V_L across the inductor is equal and opposite to V_0.

$\Rightarrow \qquad -V_0 = -L\frac{dI}{dt}$

$\Rightarrow \qquad \frac{dI}{dt} = \frac{V_0}{L}$ which is the initial (maximum) rate at which the current grows.

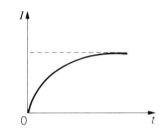

Figure 7.21

b) From Figure 7.21, after some time the current no longer grows and so the slope dI/dt is zero.

$$V_L = -L\frac{dI}{dt} = 0$$

$\Rightarrow \qquad V_0 = -(V_R + V_L) = -(V_R + 0)$

$\Rightarrow \qquad V_R = -V_0 = -IR$

$\Rightarrow \qquad IR = V_0$

The maximum current I is V_0/R

This is summarized in Table 1.

	at start	when current no longer grows
V_R	0	$-V_0$
V_L	$-V_0$	0
I	0	$\frac{V_0}{R}$

Table 1

Example 3

For the circuit shown in Figure 7.22, find
a) the maximum current
b) the maximum rate of change of current after switch S is closed.

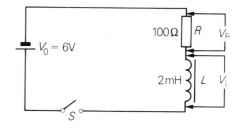

Figure 7.22

The current increases from zero to a maximum steady value resulting from the changes in p.d.

$$V_0 = -(V_L + V_R)$$

where $\quad V_L = -L\dfrac{dI}{dt} \quad$ and $\quad V_R = -IR$

a) The maximum current occurs when it reaches a steady value, i.e. when $dI/dt = 0$

$\Rightarrow \qquad V_L = 0; \quad \Rightarrow \quad V_0 = -(0 + V_R)$

$\Rightarrow \qquad V_0 = -V_R = IR$

$\Rightarrow \qquad I = \dfrac{V_0}{R} = \dfrac{6}{100} = 60 \times 10^{-3}$

The maximum current is 60 mA

b) The maximum rate of change of current is when switch S is first closed, i.e. when $I = 0$

$\Rightarrow \qquad V_R = 0; \quad \Rightarrow \quad V_0 = -(V_L + 0)$

$\Rightarrow \qquad V_0 = V_L = L\dfrac{dI}{dt}$

$\Rightarrow \qquad \dfrac{dI}{dt} = \dfrac{V_0}{L} = \dfrac{6}{2 \times 10^{-3}} = 3 \times 10^3$

The maximum rate of charge of current is $3 \times 10^3\,\text{A s}^{-1}$

Summary

When a conductor experiences a changing magnetic field, an e.m.f. is induced across it. The magnitude of the induced e.m.f. can be increased by
a) increasing the strength of magnetic field used,
b) increasing the rate of change of the magnetic field,
c) using a longer conductor.

Lenz's Law states that: the direction of the induced e.m.f. is such as to oppose the change that produces it.

If the induced e.m.f. is 1 volt when the current changes at the rate of 1 ampere per second, then the inductance is 1 henry.

$$1\,H = 1\,V\,s\,A^{-1}$$

The inductance of an inductor depends on the material of the core, the dimensions of the coil, and the number of turns on the coil.

When an inductor is connected to a d.c. supply, the current grows from zero to a maximum value; the p.d. across the inductor reduces from a maximum value to zero during this time.

The work done in building up the current in an inductor is stored in the magnetic field of the inductor. This energy stored in the magnetic field is a source of e.m.f.

The self-induced e.m.f. E in an inductor of inductance L is given by

$$E = -L\frac{dI}{dt}$$

where dI/dt is the rate at which the current changes. The self-induced e.m.f. opposes the source e.m.f.

Problems

1 Name three factors which affect the magnitude of the induced e.m.f. in a coil, explaining how each factor affects the e.m.f.

2 State Lenz's Law.

3 Describe an experiment to demonstrate mutual induction.

4 How is a self-induced e.m.f. produced? Describe an experiment to illustrate this.

5 Write down an expression for the inductance of an inductor L, giving units for quantities used.

6 List three factors which influence the inductance of an inductor. Describe an experiment which illustrates each factor.

7 Draw a current-time graph which illustrates the behaviour of the inductor in the circuit shown when switch S is closed.

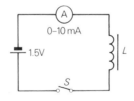

8 Describe and explain the working of a circuit which enables a 90 volt neon lamp to be operated from a 9 volt d.c. supply.

9 An inductor of 5 H is connected to a d.c. supply as shown.

When switch S is closed, a self-induced e.m.f. of 10 volts is produced. What is the maximum rate at which the current increases?

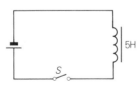

10 When the circuit shown is switched off, the current decreases at a rate of $2.0 \times 10^4\,A\,s^{-1}$. Calculate the maximum self-induced e.m.f.

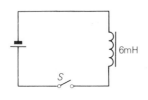

11 What value of inductor would provide a maximum self-induced e.m.f. of 100 V if the maximum rate at which the current decreases is $4.2 \times 10^4\,A\,s^{-1}$?

12 Calculate the maximum current and maximum rate of change of current in the circuit shown.

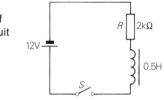

13 An inductor with an iron core is connected in series with a milliammeter to a 1.5 V supply as shown.

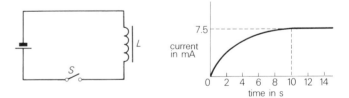

The switch S is closed and a graph is drawn to show how the current varies with time.
The graph is shown above.
a) Explain why the current takes several seconds to reach its maximum value and why it then remains constant.
b) The core is partly removed from the coil and the experiment is repeated. Sketch the graph using the same numerical scales to show how the variation of current with time differs from the first case.
                                                                    SCEEB

8 a.c. circuits

8.1 Alternating current

The electricity supplied to homes is an alternating current at an alternating voltage of 240 V having a frequency of 50 Hz. An a.c. supply is used so that a transformer can step up or step down the voltage: less energy is wasted if the electrical power is transmitted at high voltage.

A supply of alternating current is generated using an a.c. generator. The generation of alternating current in a generator is illustrated in Figure 8.1.

When a coil is rotated in a magnetic field, an electromotive force E is induced which will drive a current I through the external circuit of resistance R. At any instant the current is given by $I = E/R$.

The slip rings and carbon brushes of this generator ensure that one output terminal is always connected to one side of the coil, whether that side is moving up or down through the magnetic field as it rotates. This means that the direction of the induced electromotive force changes every half revolution of the coil.

We can use an oscilloscope to give a trace which shows how the e.m.f. varies with time. The shape of the trace is the same as that for a graph of e.m.f. against time; e.m.f. is on the vertical axis and time on the horizontal axis, Figure 8.2.

The induced e.m.f. varies between zero and a maximum value E_m known as the peak e.m.f. We can deduce how this variation takes place if we consider the coil as it rotates. Figure 8.3 shows the cross section of a coil as it rotates in a magnetic field.

The size of the induced e.m.f. is proportional to the rate at which the coil cuts through the field lines. If the resistance of the external circuit is constant, then the induced current I is proportional to the induced e.m.f. E.

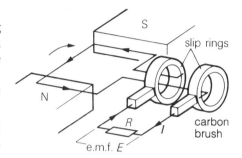

Figure 8.1 a.c. generator

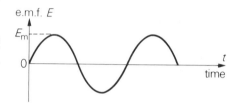

Figure 8.2 a.c. generator output

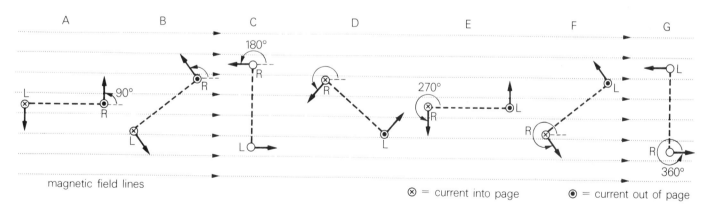

Figure 8.3 Coil rotating in a magnetic field

The direction of the induced current is shown at each stage. One side of the coil is labelled L and the other side R.
Consider the variation of current in the side of the coil marked R, Figure 8.3. When the side of the coil is moving at right angles to the magnetic field, the induced e.m.f. is a maximum. This occurs when the coil is in positions A and

E. In these two cases the currents are in opposite directions. At positions B, D and F, the induced currents are less than maximum because the loop is moving at an angle to the field. At positions C and G, the sides of the coil are moving parallel to the magnetic field and are therefore not cutting the field lines, so the induced e.m.f. is zero, Figure 8.4.

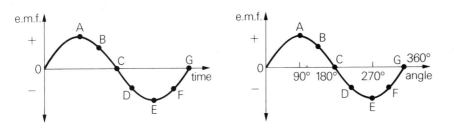

Figure 8.4 Generated e.m.f.

The pattern of e.m.f. variation is repeated at regular intervals if the generator coil rotates at a constant rate. One complete pattern is called a cycle. The time to complete one cycle is the period T which is related to the frequency f:

$$T = \frac{1}{f}$$

Electricity supplied to homes has a frequency of 50 Hz, i.e. 50 cycles per second. So the period T of voltage variation is as follows.

$$T = \frac{1}{50} = 0.02\,\text{s}$$

This is the time taken for the generator coil to rotate through 360°, i.e. 2π radians.

Figure 8.5 shows the trace obtained on a cathode ray oscilloscope of the change in generator e.m.f. E with angle θ; it is a sine curve in which the e.m.f. varies between 0 V and the peak e.m.f. E_m.

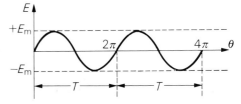

Figure 8.5

The equation is $E = E_m \sin \theta$ where θ is the angle the coil makes with the magnetic field lines.

The coil turns through 2π radians (360°) when one cycle is completed, i.e. in time T. So the angle θ turned in time t is given by

$$\frac{\theta}{t} = \frac{2\pi}{T}$$

$$\Rightarrow \theta = \frac{2\pi t}{T} = 2\pi f t \text{ because } T = \frac{1}{f}$$

$$\Rightarrow E = E_m \sin 2\pi f t$$

This equation gives the instantaneous e.m.f. E, at time t, in terms of the maximum e.m.f. E_m and the frequency f.

In the same way, the instantaneous current I is given in terms of the maximum current I_m:

$$I = I_m \sin 2\pi f t$$

Comparison of a.c. and d.c.

The current and the p.d. of an a.c. supply vary continuously and so the effective value of these quantities is less than the maximum value.

The circuit in Figure 8.6 can be used to compare the heating effect of a d.c. supply with that of an a.c. supply. A light meter is placed to measure the output of lamp B first from a low voltage a.c. supply and then from a d.c. supply. Resistor R is adjusted until the lamp is equally bright from both sources. The peak voltage V_m of the a.c. supply and the equivalent steady voltage V_{dc} of the d.c. supply are measured on the calibrated screen of a C.R.O., Figure 8.7.

Typical results for three different lamps are shown in Table 1 below.

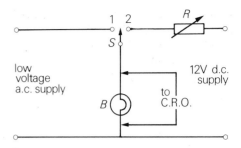

Figure 8.6 Comparison of a.c. and d.c. supplies

	V_m (volts)	V_{dc} (volts)	V_{dc}/V_m
lamp 1	12.0	8.5	0.7
lamp 2	3.0	2.1	0.7
lamp 3	2.0	1.4	0.7

Table 1

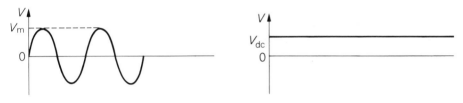

Figure 8.7

These show that the equivalent d.c. voltage V_{dc} for the same power output is about 70% of the maximum a.c. voltage V_m.

$$V_{dc} = 0.7\,V_m$$

Mathematical treatment

The above experiment compared the power outputs of an a.c. and a d.c. supply. The power P released from a resistor is as follows.

$$P = V \times I$$
$$= \frac{V^2}{R} \text{ where } R \text{ is the value of the resistor}$$

For a d.c. supply $P = V_{dc}^2/R$

For an a.c. supply $P = V_{ac}^2/R$

Since the two power outputs are the same

$$V_{dc}^2/R = V_{ac}^2/R$$
$$\Rightarrow \quad V_{dc}^2 = V_{ac}^2$$
$$\text{and} \quad V_{ac}^2 = (V_m \sin 2\pi f t)^2$$
$$= V_m^2 \sin^2 2\pi f t$$

Figure 8.8 compares the graph of V_{ac} against t with that of V_{ac}^2 against t. The curve in the second graph fluctuates also but is always positive; the average value of this curve is $\frac{1}{2}V_m^2$ because the curve is symmetrical about this value.

$$\Rightarrow \quad V_{dc}^2 = \tfrac{1}{2}\,V_m^{\,2}$$
$$\Rightarrow \quad V_{dc} = \frac{V_m}{\sqrt{2}} \approx 0.7\,V_m$$

This average value is known as the **root mean square voltage** V_{rms}.

$$V_{rms} = 0.7\,V_m$$

Similarly the root mean square current I_{rms} is $0.7\,I_m$.

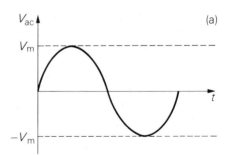

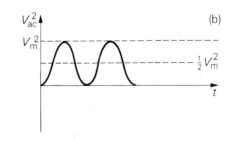

Figure 8.8

Example 1

Calculate the peak voltage of a 240 V a.c. supply.

$$V_{rms} = 0.7 V_m$$

where $V_{rms} = 240$ V and V_m is the peak voltage

$$\Rightarrow \quad V_m = \frac{V_{rms}}{0.7}$$

$$= \frac{240}{0.7} \approx 339$$

The peak voltage of a 240 V a.c. supply is 339 V

Example 2

The peak value of an alternating current in a $10 \, \Omega$ resistor is 3.0 A. Calculate the power developed in the resistor.

$$I_{rms} = 0.7 I_m$$

where $\quad I_m = 3.0$ A

$$\Rightarrow \quad I_{rms} = 0.7 \times 3.0 = 2.1$$

$$\text{power developed} \quad = I_{rms}^2 \times R$$

$$= 2.1^2 \times 10$$

$$= 44$$

The resistor develops a power of 44 W

8.2   Resistors, capacitors and inductors in a.c. circuits

When capacitors and inductors are connected to a d.c. supply, their p.d.'s and currents do not reach their maximum values at the same instant: delays occur. The observation of these delays while we are using a varying, a.c. supply is difficult unless we reduce the supply frequency to about 1 Hz. This can be achieved using a slow a.c. generator of the type shown in Figure 8.9, or by using the very low frequency output from a variable frequency signal generator.

Figure 8.9 Slow a.c. generator

Phase difference

Using the circuit in Figure 8.10 it is possible to observe the potential difference and current variation when a **resistor** is supplied with low frequency a.c.

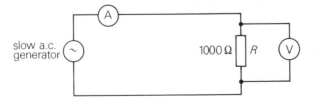

Figure 8.10 Phase in a resistor circuit

There is **no** difference in phase between the voltmeter readings and the ammeter readings as shown in Figure 8.11.

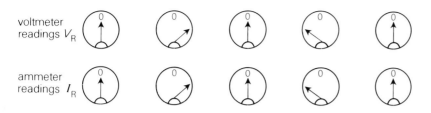

voltmeter
readings V_R

ammeter
readings I_R

Figure 8.11

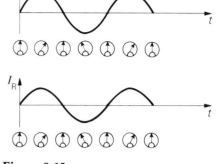

Figure 8.12

The current and voltage variations match, and we say that the current and voltage are **in phase**. This can be represented graphically as in Figure 8.12.

The current I_R and voltage V_R reach their maximum values at the same instant.

If the resistor in Figure 8.10 is replaced by a 200 μF **capacitor** we obtain a different set of results, Figure 8.13.

slow a.c.
generator   200 μF C

Figure 8.13 Phase in a capacitor circuit

There is a **difference** in phase between the voltmeter readings and the ammeter readings as shown in Figure 8.14.

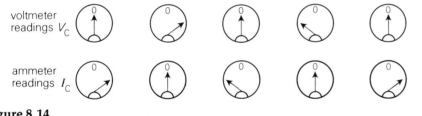

voltmeter
readings V_C

ammeter
readings I_C

Figure 8.14

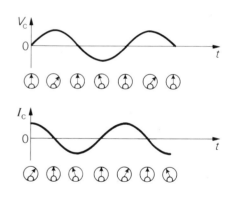

Figure 8.15

The current and voltage variations do not match and we say that the current and voltage are **out of phase**. This is shown graphically in Figure 8.15.

The current I_c reaches its maximum value before the voltage V_c reaches its maximum value. **The current leads the voltage.**

If we now replace the capacitor by a 2400 turn iron-cored **inductor** the results are different again, Figure 8.16.

slow a.c.
generator   2400 turns L

Figure 8.16 Phase in an inductor circuit

There is a **difference** in phase between the voltmeter readings and the ammeter readings as shown in Figure 8.17 on the next page.

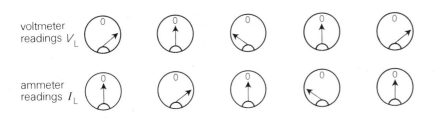

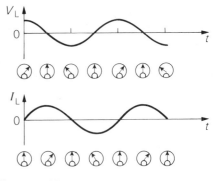

Figure 8.17

Figure 8.18

The voltage and current readings do not match and we say that the voltage and current are **out of phase**. This is illustrated graphically in Figure 8.18.

The current I_L reaches its maximum value after the voltage V_L has reached its maximum value. In this case, the **voltage leads the current.**

Measuring the opposition

The opposition of a component to alternating current in a circuit is called its **impedance**. Impedance is measured in ohms and is defined as the ratio of the r.m.s. potential difference to the r.m.s. current for the component in question. In fact, it is the potential difference across the component divided by the current in the component. When dealing with a.c. circuits, we use a.c. meters to give r.m.s. values.

There are three types of impedance with symbols R, X_c and X_L.

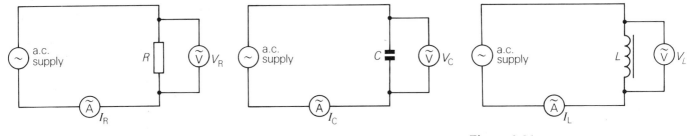

Figure 8.19 **Figure 8.20** **Figure 8.21**

The impedance of a resistor is the same for both a.c. and d.c. and is its resistance R. For the resistor in Figure 8.19.

$$R = \frac{V_R}{I_R} \quad \text{where } V_R = \text{r.m.s. voltage across } R$$
$$I_R = \text{r.m.s. current in } R$$

The impedance of a capacitor is its capacitive reactance X_c. For the capacitor in Figure 8.20

$$X_c = \frac{V_c}{I_c} \quad \text{where } V_c = \text{r.m.s. voltage across } C$$
$$I_c = \text{r.m.s. current}$$

The impedance of an inductor is its inductive reactance X_L. For the inductor in Figure 8.21

$$X_L = \frac{V_L}{I_R} \quad \text{where } V_L = \text{r.m.s. voltage across } L$$
$$I_L = \text{r.m.s. current in } L$$

a) The impedance of a resistor (its resistance) can be calculated from measurements on the circuit shown in Figure 8.22.

For a given frequency of a.c. supply, the current and the potential difference are measured when the switch is closed, using a.c. meters.

Sample results: $I_R = 0.2\,\text{A}$; $V_R = 2.0\,\text{V}$

$$R = \frac{V_R}{I_R} = \frac{2.0}{0.2} = 10\,\Omega$$

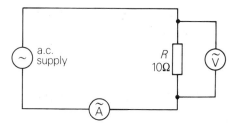

Figure 8.22 Measuring resistance

b) The reactance of a capacitor can similarly be calculated for a given frequency of a.c. supply from measurements on the circuit shown in Figure 8.23.

Sample results: $I_c = 1.36\,\text{A}$; $V_c = 1.0\,\text{V}$

$$X_c = \frac{V_c}{I_c} = \frac{1.0}{1.36} = 0.74\,\Omega$$

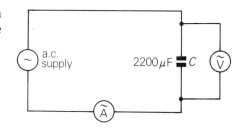

Figure 8.23 Measuring capacitive reactance

c) The reactance of an inductor can likewise be calculated for a given frequency of a.c. supply from measurements on the circuit shown in Figure 8.24.

Sample results: $I_L = 0.0064\,\text{A}$; $V_L = 2.0\,\text{V}$

$$X_L = \frac{V_L}{I_L} = \frac{2.0}{0.0064} = 312\,\Omega$$

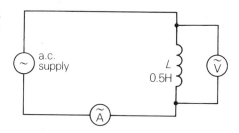

Figure 8.24 Measuring inductive reactance.

Example 3

Calculate the reactance of the capacitor in the circuit shown in Figure 8.25.

$$I_c = 10\,\text{mA}; \quad V_c = 2.6\,\text{V}$$

$$X_c = \frac{V_c}{I_c}$$

$$= \frac{2.6}{10 \times 10^{-3}} = 260$$

The capacitive reactance is 260 ohms

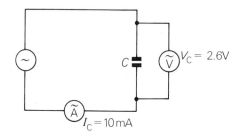

Figure 8.25

Example 4

In Figure 8.26, the reactance of the inductor is 300 ohms when the peak p.d. across it is 12 volts. Assuming the inductor has zero resistance, calculate the value of the peak current in the circuit.

$$X_L = 300\,\Omega; \quad V_L = 12\,\text{V}$$

$$X_L = \frac{V_L}{I_L}$$

$$\Rightarrow 300 = \frac{12}{I_L}$$

$$\Rightarrow I_L = \frac{12}{300} = 0.04$$

The peak current is 40 mA

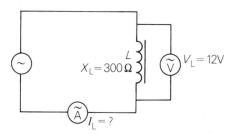

Figure 8.26

Factors affecting capacitive reactance

a) Frequency

The instantaneous current I in a capacitor circuit depends on the rate at which the voltage changes (page 69)

$$I_c = C\frac{dV}{dt}$$

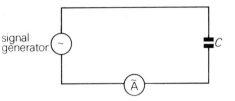

Figure 8.27

At high frequency, the voltage changes very rapidly and so the current is large. But $X_c = V_c/I_c$; so if the supply voltage is constant, the reactance X_c is small when the current is large at high frequency.

The circuit in Figure 8.27 is used to investigate how the current I_c depends on the supply frequency. Table 2 shows the results obtained for a constant supply voltage, and these are plotted on the graph in Figure 8.28.

frequency f (Hz)	1000	2000	3000
current I_c (A)	0.025	0.05	0.075

Table 2

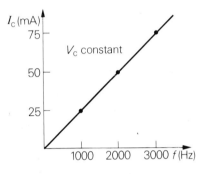

Figure 8.28

A straight line through the origin is obtained showing that the current I_c varies directly as the frequency.

$$I_c \propto f$$

But V_c is constant, and $X_c = V_c/I_c$ $\Rightarrow$ $X_c \propto \frac{1}{f}$ ...[1]

b) Capacitance

The same circuit can be used to investigate how the current I_c depends on the capacitance C. Table 3 shows the results obtained for a constant supply voltage and signal generator frequency when various capacitors are inserted in the circuit. The results are plotted on the graph in Figure 8.29.

capacitance C (μF)	1000	2000	3000
current I_c (A)	0.03	0.06	0.09

Table 3

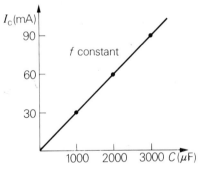

Figure 8.29

The graph shows that the current I_c varies directly as the capacitance.

$$I_c \propto C$$

Again V_c is constant and $X_c = V_c/I_c$ $\Rightarrow$ $X_c \propto \frac{1}{C}$ ...[2]

Combining equations [1] and [2], the capacitive reactance is given by

$$X_c \propto \frac{1}{fC} \Rightarrow X_c = \frac{k}{fC}$$

The value of k is $1/2\pi$ $\Rightarrow$ $X_c = \frac{1}{2\pi fC}$ ...[3]

Example 5

Calculate the capacitive reactance of a $100\,\mu$F capacitor in a circuit supplied with 12 V alternating current at a frequency of 50 Hz.

$$X_c = \frac{1}{2\pi fC} \quad \text{where } f = 50\,\text{Hz} \quad C = 100\,\mu\text{F}$$

$$\Rightarrow X_c = \frac{1}{2\pi \times 50 \times 100 \times 10^{-6}} = 32$$

The capacitive reactance is 32 Ω

Example 6

Calculate the current in a circuit consisting of a 2000 μF capacitor connected to a 20 V supply at a frequency of 25 Hz.

$$X_c = \frac{1}{2\pi f C} \quad \text{where } f = 25 \text{ Hz} \quad C = 2000 \, \mu\text{F} \quad V_c = 20 \text{ V}$$

$$\Rightarrow X_c = \frac{1}{2\pi \times 25 \times 2000 \times 10^{-6}} = 3.2$$

$$I_c = \frac{V_c}{X_c} = \frac{20}{3.2} = 6.3$$

The current is 6.3 A

Factors affecting inductive reactance

Using the circuit in Figure 8.30, similar experiments to those with a capacitor can be conducted on an inductor to find the relationship between
    a) the inductive reactance X_L and the supply frequency f
and **b)** the inductive reactance X_L and the inductance L.
    The three equations below assume that the electrical resistance of the wire in the inductor is small enough to be neglected.
    For any inductor when V_L is constant, the current I_L decreases as the frequency increases (Figure 8.31), so the reactance X_L also increases.

$$X_L \propto f \quad ...[4]$$

For a range of inductors when f is constant

$$X_L \propto L \quad ...[5]$$

These two equations are combined to give

$$X_L = 2\pi f L \quad ...[6]$$

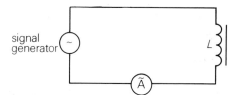

Figure 8.30

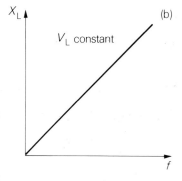

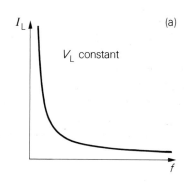

Figure 8.31

Example 7

Calculate the inductive reactance of a 0.50 H inductor in a circuit supplied with 12 V alternating current at a frequency of 50 Hz.

$$X_L = 2\pi f L \quad \text{where } f = 50 \text{ Hz} \quad L = 0.50 \text{ H}$$
$$\Rightarrow X_L = 2\pi \times 50 \times 0.50 = 157$$

The inductive reactance is 157 Ω

Example 8

Find the voltage if the current through a 0.50 H inductor is 1.0 mA and the supply frequency is 120 Hz.

$$X_L = 2\pi f L \quad \text{where } f = 120 \text{ Hz} \quad L = 0.50 \text{ H}$$

$$\Rightarrow X_L = 2\pi \times 120 \times 0.50 = 377 \, \Omega$$

$$X_L = \frac{V_L}{I_L} \quad \text{where } V_L = ? \quad I_L = 10 \text{ mA}$$

$$\Rightarrow V_L = X_L \times I_L$$
$$= 377 \times 10 \times 10^{-3} = 3.77$$

The voltage is 3.8 V

8.3 Capacitors and inductors combined

Table 5 summarizes how the reactance of a capacitor or an inductor is affected by the frequency of the supply.

frequency	X_c	X_L
low	large	small
high	small	large

Table 5

We shall now look at the combination of an inductor and a capacitor in the same circuit, and how the resultant impedance of the circuit changes with the frequency of the supply.

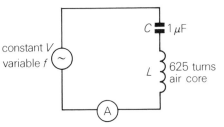

Figure 8.32

L and C in series

Figure 8.32 shows a signal generator supplying a constant voltage at a variable frequency to an inductor and a capacitor in **series**. The ammeter readings are shown for various supply frequencies in the graph, Figure 8.33.

It is seen that the average current I in the circuit changes sharply as the frequency is varied, and that it reaches a peak at a frequency f_0.

From Table 5 we know that the capacitive reactance X_c decreases with increasing frequency, and that the inductive reactance X_L increases with increasing frequency. With increasing frequency therefore, I_c is reduced from a high value but I_L is increased from a low value. At some value f_0 of the frequency, these two quantities are equal to give a maximum value for the average current I. The frequency f_0 is known as the **resonant frequency**.

Variation of the impedance X of the circuit resulting from a combination of X_c and X_L is shown in the graph of Figure 8.34. The resonant frequency f_0 is at the point where X_L equals X_c to give a maximum current.

The value of f_0 is found as follows. At this frequency, $X_L = X_c$;

but $X_L = 2\pi f L$ and $X_c = \dfrac{1}{2\pi f C}$

$$\Rightarrow 2\pi f_0 L = \frac{1}{2\pi f_0 C}$$

$$\Rightarrow f_0^2 = \frac{1}{(2\pi L)(2\pi C)}$$

$$\Rightarrow f_0 = \frac{1}{2\pi \sqrt{(LC)}}$$

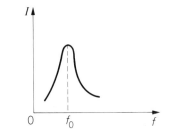

Figure 8.33 Graph of I against f for L and C in series

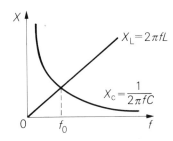

Figure 8.34 Variation of impedance with frequency for L and C in series

Example 9

The circuit in Figure 8.35 includes an inductor and a capacitor in series. Calculate the resonant frequency which gives a maximum supply current.

$$f_0 = \frac{1}{2\pi \sqrt{(LC)}} \quad \text{where } C = 100\,\mu\text{F} \quad L = 0.50\,\text{H}$$

$$\Rightarrow f_0 = \frac{1}{2\pi \sqrt{(0.50 \times 100 \times 10^{-6})}} = 22.5$$

The resonant frequency is 23 Hz

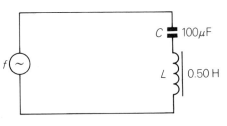

Figure 8.35

L and *C* in parallel

Figure 8.36 shows a signal generator supplying a constant voltage at a variable frequency to an inductor and a capacitor in **parallel**.

As before, at low frequency X_c is large and X_L is small and so most of the current is routed through the inductor *L*, Figure 8.37. Conversely, at high frequency most of the current is routed through the capacitor *C*, Figure 8.38, because X_c is small and X_L is large.

The graph in Figure 8.39 shows how the average current *I* in the circuit again changes sharply as the frequency is varied, but for this circuit it reaches a **minimum** at a frequency f_0.

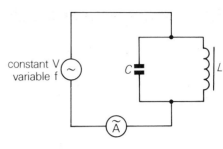

Figure 8.36

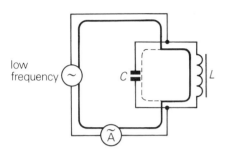

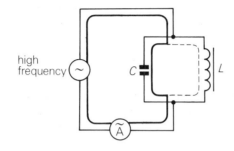

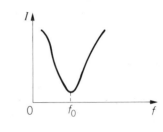

Figure 8.37 Low frequency route of current

Figure 8.38 High frequency route of current

Figure 8.39 Graph of *I* against *f* for *L* and *C* in parallel

At the resonant frequency f_0, the current I_c through the capacitor equals the current I_L through the inductor in the reverse direction: the currents are exactly out of phase if the inductor has negligible resistance. So the supply current is a minimum.

The value of f_0 is found as before from the same formula.

$$X_L = X_c$$

$$\Rightarrow 2\pi f_0 L = \frac{1}{2\pi f C}$$

$$\Rightarrow \quad f_0 = \frac{1}{2\pi \sqrt{(L C)}}$$

Example 10

Figure 8.40 shows a circuit in which an inductor and a capacitor are in parallel. Calculate the frequency at which the supply current is a minimum.

$$f_0 = \frac{1}{2\pi \sqrt{(LC)}} \quad \text{where } C = 2000\,\mu F \quad L = 10\,mH$$

$$\Rightarrow f_0 = \frac{1}{2\pi \sqrt{(10 \times 10^{-3} \times 2000 \times 10^{-6})}} = 35.5$$

For minimum current, the frequency is 36 Hz

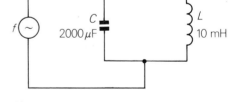

Figure 8.40

8.4   Some uses of capacitors and inductors

Capacitors

The reactance X_c of a capacitor is $\frac{1}{2\pi f C}$.

A d.c. supply has zero frequency, and so X_c is then infinite. Thus a capacitor will not pass a d.c. signal.

Figure 8.41(a) shows the graph of an electrical signal that has both an a.c. component V_{ac} and a d.c. component V_{dc}. When this signal is supplied to the circuit shown in Figure 8.42, the voltage variation V_R across resistor R no longer includes the d.c. component, Figure 8.41(b). The capacitor has blocked the d.c. signal.

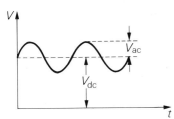

(a) signal containing a.c. and d.c.

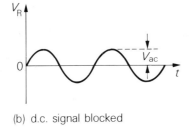

(b) d.c. signal blocked

Figure 8.41

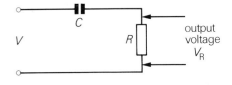

Figure 8.42

A capacitor can also be used to smooth rectified a.c. to give d.c. The circuit in Figure 8.43 can be used to demonstrate the smoothing action of a capacitor C_1. Without C_1, the variation in output voltage would be as shown in Figure 8.44(a). When capacitor C_1 is included in the circuit, the output is smoothed to the form shown in Figure 8.44(b). The smoothing action is due to the charging and discharging of capacitor C_1. This output is called a ripple voltage.

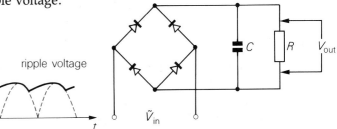

Figure 8.43

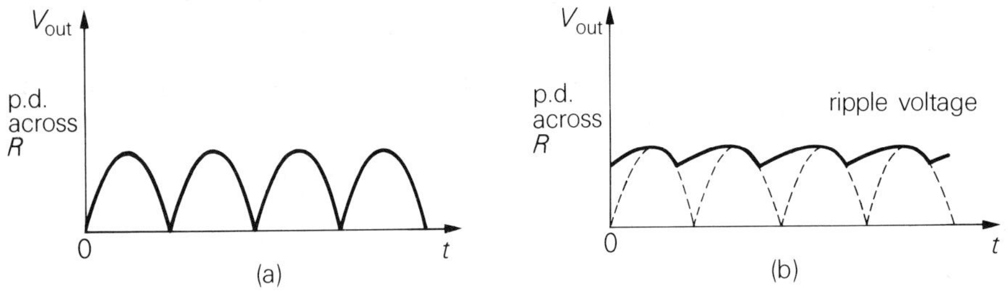

(a)

(b)

Figure 8.44 Capacitor smoothing

Inductors

An inductor may be used to further smooth rectified a.c. as shown in the circuit of Figure 8.45.

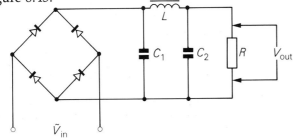

Figure 8.45

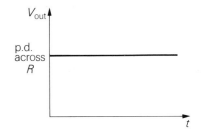

Figure 8.46

The addition of an extra capacitor C_2 and an inductor L provides a steady d.c. output voltage across R, Figure 8.46. The inductor L has a high impedance to the ripple voltage of Figure 8.44(b). An iron-cored inductor used in this way is called a **choke**.

The circuit shown in Figure 8.47 is for lighting a fluorescent tube. The choke is used to create a high voltage surge that is required to start the discharge in the tube when S_2 switches off. This inductor then restricts the current while the tube is lighting up.

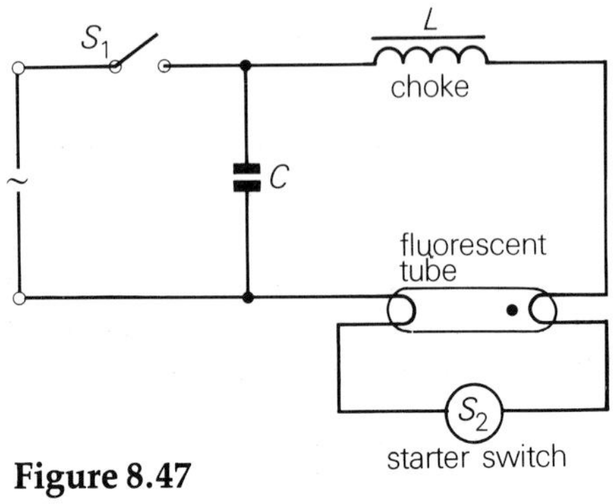

Figure 8.47

Summary

An alternating e.m.f. may be described by the equation
$$E = E_m \sin 2\pi ft$$
where E_m is the peak or maximum e.m.f.; f is the frequency and t is time.

An alternating current may be described by the equation
$$I = I_m \sin 2\pi ft$$
where I_m is the peak or maximum current.

V_{rms} is the root mean square voltage and represents the d.c. equivalent of an a.c. voltage variation
$$V_{rms} = \frac{V_m}{\sqrt{2}} = 0.7\,V_m$$

When a resistor is inserted in an a.c. circuit, the current and p.d. variation are in phase.

When a capacitor is inserted in an a.c. circuit, the current and voltage across the capacitor are out of phase. The current reaches its maximum value before the voltage reaches its maximum value. The current leads the voltage.

When an inductor is inserted in an a.c. circuit, the voltage and current across the inductor are out of phase. The voltage reaches its maximum value before the current reaches its maximum value. The voltage leads the current.

In an a.c. circuit

resistance $R = \dfrac{V_R}{I_R}$ where V_R is r.m.s. voltage across R and I_R is r.m.s. current in R

capacitive reactance $X_c = \dfrac{V_c}{I_c}$ where V_c is r.m.s. voltage across C and I_c is r.m.s. current

inductive reactance $X_L = \dfrac{V_L}{I_L}$ where V_L is r.m.s. voltage across L and I_L is r.m.s. current in L

Capacitive reactance can be calculated from
$$X_c = \frac{1}{2\pi fC} \text{ ohms}$$

Inductive reactance can be calculated from
$$X_L = 2\pi fL \text{ ohms}$$

At the resonant frequency f_0, the inductive reactance and the capacitive reactance are equal. The resonant frequency can be calculated from
$$f_0 = \frac{1}{2\pi\sqrt{(LC)}}$$

Problems

1 Explain the symbols in the equation
$E = E_m \sin 2\pi ft$

2 Describe an experiment to determine the relationship between peak voltage and root mean square voltage.

3 What is the peak voltage of a 12 V r.m.s. supply?

4 The peak value of an alternating current in a 3 Ω lamp filament is 4 A. Calculate the power of the lamp.

5 In the diagram, the two lamps B_1 and B_2 are lit by a variable frequency a.c. supply.

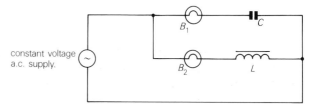

If the frequency of the supply is initially at 60 Hz and then increased to 12 kHz while the voltage is kept constant, what happens to the brightness of each lamp?

6 The readings given in the table were obtained using the circuit shown.

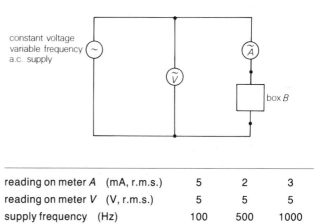

reading on meter A (mA, r.m.s.)	5	2	3
reading on meter V (V, r.m.s.)	5	5	5
supply frequency (Hz)	100	500	1000

Suggest what circuit box B might contain.

7 Calculate the impedance of each component in the circuit below, and hence indicate the order of brightness of lamps B_1, B_2 and B_3.

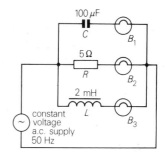

97

8 Calculate the resonant frequency for the following circuit.

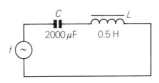

9 What value of inductor is required for the following circuit to resonate at 50 MHz?

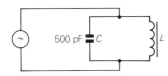

10 a) Give a circuit diagram of the apparatus you would use to show that the capacitive reactance of a capacitor changes as the frequency of the a.c. supply is changed.
Describe how you would use the apparatus and state what you would expect to find.
b) The following results were obtained in an experiment to find how capacitive reactance is related to capacitance.

Capacitance C (μF)	Capacitive Reactance X_c (ohms)
4	797
8	398
16	200
25	128

i) Use a graphical method to establish a relationship between the capacitance C and the capacitive reactance X_c.
ii) Explain why the frequency of the source must be kept constant during the experiment.

SCEEB

11 A capacitor is connected in series with a resistor, an inductor and a signal generator which provides a constant voltage at whatever frequency it is used.

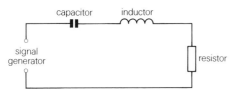

Describe how the resonant frequency of this circuit may be found using
i) an a.c. ammeter only
ii) a cathode ray oscilloscope only.

SCEEB

12 An inductor and a capacitor are connected between two terminals inside each of two sealed boxes. In one box they are connected in series while in the other they are connected in parallel.
Explain how you would use an a.c. ammeter and a signal generator to identify which circuit was connected inside each box.

SCEEB

13 a) The diagram shows an inductor connected in series with an ammeter and signal generator. The output voltage of the signal generator is kept constant.

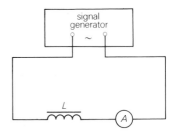

The current is noted at various frequencies and a current-frequency graph is drawn.
The procedure is repeated using a capacitor instead of the inductor.
Both graphs are shown below.

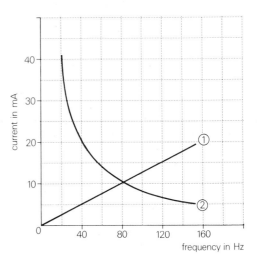

State and explain which graph was produced using the inductor.
b) i) Describe with the aid of a circuit diagram how the resonant frequency of the capacitor and inductor when connected in parallel may be found.
ii) State the resonant frequency in this case.

SCEEB

14 A signal generator has a fixed voltage, variable frequency output. It is connected to an inductor L and a capacitor C along with two suitable voltmeters, V_1 and V_2. The circuit resonates at 250 Hz.

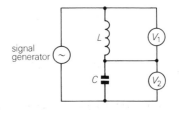

a) Compare the readings on V_1 and V_2 as the supply frequency is increased from 50 Hz to 500 Hz.
b) Sketch a graph of supply current against frequency for the above circuit in the range 50 Hz to 500 Hz.

15   The circuit shows a capacitor, an inductor and a resistor, connected in series with a signal generator which delivers a fixed value of alternating voltage.

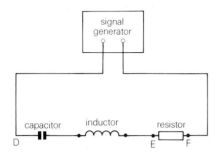

a) The Y plates of an oscilloscope, which has its time-base switched off, are connected **in turn** across (i) EF; (ii) DE.
In **each** case describe how the trace on the oscilloscope screen changes as the frequency of the supply voltage is increased through the resonant frequency of the circuit.
b) The time-base of the oscilloscope is switched on and set to $2\,\text{ms}\,\text{cm}^{-1}$. The Y plates of the oscilloscope are now connected across DF. At the circuit resonant frequency, 4 cycles of the signal appear on the screen which is 10 cm wide. Calculate the value of this resonant frequency.
c) The resonant frequency f for a circuit like this depends on a number of factors, one of which is the inductance value L.
A group of girls decides to investigate this and obtains the following set of results for the resonant frequency of the circuit for various inductance values.

L (in henries)	0.5	1.0	2.0	4.0	8.0
f (in hertz)	200	141	100	70.0	50.5

(The henry is a unit of inductance.)
The girls suggest that in order to determine the numerical relationship between L and f, values should be calculated for

$$f \times L; \quad \frac{f}{L}; \quad f^2 \times L; \quad \frac{f^2}{L}.$$

i) By **inspecting** the data, state which of these suggestions you would reject. Give your reasoning.
ii) Hence, or otherwise, determine the numerical relationship between L and f, showing clearly the steps involved.

SCEEB

16   When the output of an audio amplifier is connected across AB in the circuit shown, one loudspeaker emits high frequency sound waves while the other emits low frequency sound waves.

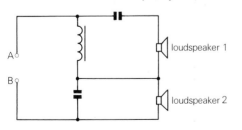

State which loudspeaker emits the low frequency sound waves. Explain fully how the circuit components direct low frequency sounds to this loudspeaker and high frequency sounds to the other one.

SCEEB

9 Geometrical optics

9.1 How do we see?

When we see an object, we do so because light from that object enters our eyes. It may be that the object emits its own light (e.g. a candle, lamp or star), in which case the object is said to be 'luminous'. Most of the objects that we see do not emit their own light and we see them because they reflect light. In this case the objects are said to be 'illuminated'.

When we consider how we see objects, it is useful to draw 'rays' which are straight lines showing the directions in which light travels. A ray diagram can be considered as a diagram showing the paths which would be taken by narrow beams of light.

If we view an object, light from all visible parts of the object enters our eyes. It helps to understand how an object is seen and located if we consider light from one point on the object only. Figure 9.1 shows rays drawn from a point on the surface of an object. Some of the rays are shown entering the eye. Until we have studied the properties of lenses, we cannot say what happens to the light when it enters the eye but we can say that, to the observer, the point is seen as being the source of this cone of light entering the eye.

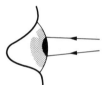

(a) Rays entering the eye from a point

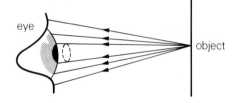

Figure 9.1 Rays from a point on an object

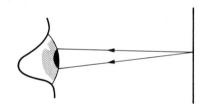

Figure 9.2 Two-dimensional ray diagram

If light from a point enters the eye, as shown in Figure 9.3(a), the observer will see the point as if it were at O as shown in Figure 9.3(b). As we shall see, these rays may have been bent (refracted) or reflected, and may not actually come from O, but the position at which the observer sees the point can be located by projecting back the rays entering the eye to find the point at which they meet.

For simplicity, most rays diagrams are shown in two dimensions, and Figure 9.2 shows how we represent the cone of light that enters the eye from the point on the object.

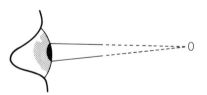

(b) Apparent source of rays from a point

Figure 9.3

Locating an Image

When you look in a mirror, you see a picture 'behind' the mirror. You know that what you are seeing is not actually behind the mirror but that it is a picture formed by light from objects in front of the mirror being reflected by the mirror. Such a 'picture', where the eyes sees objects as being in some place which is different from the actual object, is called an 'image'. When you look in a mirror, you see an image of the scene in front of it, Figure 9.4.

Figure 9.4

A method of locating an image can be demonstrated by a simple experiment. Hold one finger directly above another, Figure 9.5(a), and move your head from side to side. Not surprisingly, the two fingers always appear one directly above the other.

Now hold two fingers in line, with one above the other but with the lower one nearer to your eye, Figure 9.5(b). They appear together when viewed head-on but, when you move your head from side to side, they will appear to separate. The plan diagram in Figure 9.6 shows how this occurs.

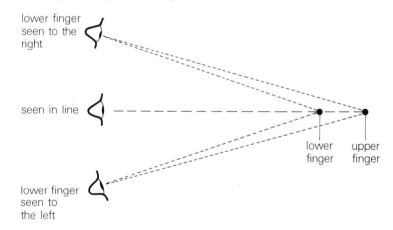

Figure 9.6

Thus the two fingers are only seen together at all times when they are actually together; at other times they are seen to separate when the observer moves his head from side to side. If something can be placed in a position so that it coincides with an image, the object and the image will be seen together when the observer moves his head from side to side. This fact can be used in an experiment to find the relationship between the position of an object and its image in a mirror.

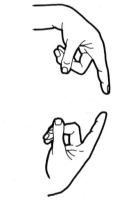

(a) one directly above the other

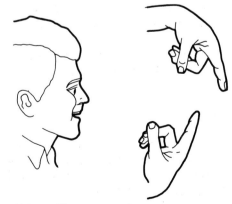

(b) lower finger nearer the eye

Figure 9.5

9.2 Reflection by a plane mirror

The following experiment is designed to investigate the relationship between the position of an object and its image in a plane (flat) mirror. The plane mirror is set up as shown in Figure 9.7, and the observer sees the image of the candle in the mirror.

The observer then moves a second candle behind the mirror until it coincides with the image of the first candle, Figure 9.8. He knows that the image and the second candle are at the same position when the movement of his head from side to side does not separate the image from the candle.

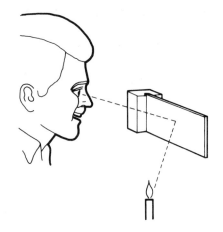

Figure 9.7 Reflection by a plane mirror

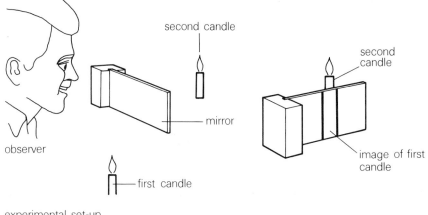

Figure 9.8 Locating the image

Figure 9.9 View seen by experimenter

If the positions of the two candles and the mirror are then marked, it is found that both candles are at equal distances from the reflecting surface of the mirror, and that a line joining the two candle positions is perpendicular to the mirror surface. Since the second candle was placed to coincide with the image of the first candle, we can conclude that, when the reflection of an object is viewed in a plane mirror,

a) the object and its image are at equal perpendicular distances from the reflecting surface of the mirror;

b) the line joining the object to its image is perpendicular to the surface of the mirror, Figure 9.10. A line perpendicular to the surface of the mirror is called a normal.

I is the image v = image distance

O is the object u = object distance

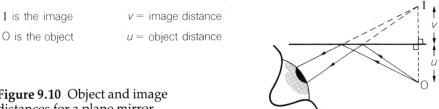

Figure 9.10 Object and image distances for a plane mirror

The observer sees the rays as coming from I behind the mirror, but they do not actually do so; for this reason, the image is called a virtual (or apparent) image.

When the image is formed at a position from which the rays **appear** to come rather than at a position from which the rays actually come, the image is called a **virtual** image.

9.3 Laws of reflection

When a beam of light is reflected, it obeys the Laws of Reflection as follows.

1 The incident ray, reflected ray and the normal at the point of reflection are all in the same plane, Figure 9.11.

2 The angle of incidence equals the angle of reflection, Figure 9.12.

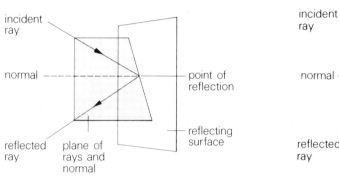

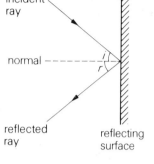

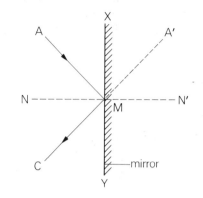

Figure 9.11 First law of reflection

Figure 9.12 Second law of reflection

Figure 9.13 Single ray reflected

Applying the Laws of Reflection to find the position of an image in a plane mirror

Figure 9.13 shows a single ray of light reflected by a plane mirror. MA' is an extension of the path CM of the reflected ray.

$\hat{AMN} = \hat{CMN}$ (Law of Reflection)
$\hat{A'MN'} = \hat{CMN}$ (Opposite angles of intersecting straight lines)

$\hat{XMA'} = 90° - \hat{A'MN'}$
$\hat{XMA} = 90° - \hat{AMN}$ $\Rightarrow$ $\hat{XMA'} = \hat{XMA}$

$\hat{YMA} = 180° - \hat{XMA}$
$\hat{YMA'} = 180° - \hat{XMA'}$ $\Rightarrow$ $\hat{YMA} = \hat{YMA'}$

Figure 9.14 shows two rays from a point O on an object, reflected by a plane mirror. To an observer, the rays appear to come from I; therefore the point I is the image of point O.

Equal angles are marked in for the two rays. This shows that, in triangles OXY and IXY

$\hat{OXY} = \hat{IXY}$ $\hat{OYX} = \hat{IYX}$ XY is a side common to both triangles.

$\Rightarrow$ triangles OXY and IXY are congruent.

Since the perpendicular distance of I from the mirror is the altitude of the triangle IXY and the perpendicular distance of O from the mirror is the altitude of triangle OXY, these distances are equal.

Thus the distance IM of the image from the mirror equals the distance OM of the object from the mirror, Figure 9.15.

The distance of the image from the plane mirror always equals the distance of the object from the plane mirror. This agrees with the result of the experiment in which the image of a candle was found to be as far behind the mirror as the candle was in front.

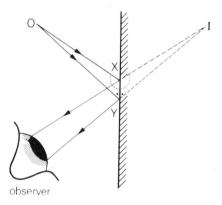

Figure 9.14 Image seen in plane mirror

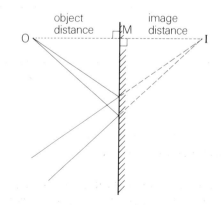

Figure 9.15 Object and image distances

9.4   Refraction of light

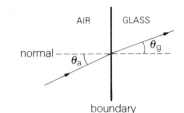

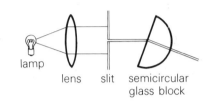

Figure 9.16 Refraction of light

When a ray of light passes from air to glass, it is bent towards the normal so that the angle θ_a between the normal and the ray in air is greater than the angle θ_g between the normal and the ray in glass, Figure 9.16.

The apparatus shown in Figure 9.17 is used to investigate the relationship between θ_a and θ_g. Since the glass block is semicircular, any beam of light through the centre of the straight face of the block travels along a radius in the glass and will therefore be normal to the curved face. The lens is used to produce a parallel beam of light and the slit is used to produce a narrow beam. By varying the angle between the ray and the glass block, a series of measurements of θ_a and θ_g is made. Table 1 shows a typical set of such measurements. If these results are plotted on a graph, the graph shown in Figure 9.18 is obtained.

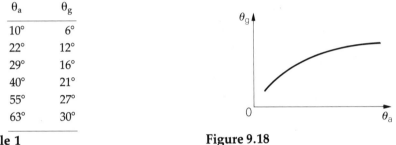

Figure 9.17 Investigation of relationship between θ_a and θ_g

θ_a	θ_g
10°	6°
22°	12°
29°	16°
40°	21°
55°	27°
63°	30°

Table 1

Figure 9.18

Neither the table of results nor the graph shows any obvious mathematical relationship between θ_a and θ_g. However, if we take the sines of the angles (Table 2) and plot a graph of $\sin \theta_a$ against $\sin \theta_g$, we obtain the graph shown in Figure 9.19.

θ_a	θ_g	$\sin \theta_a$	$\sin \theta_g$
10°	6°	0.17	0.10
22°	12°	0.37	0.21
29°	16°	0.48	0.28
40°	21°	0.64	0.36
55°	27°	0.82	0.45
63°	30°	0.89	0.50

Table 2

Figure 9.19

Since the graph of $\sin \theta_a$ against $\sin \theta_g$ is a straight line passing through the origin,

$$\sin \theta_a \propto \sin \theta_g$$
$$\Rightarrow \sin \theta_a = \text{constant} \times \sin \theta_g$$
$$\Rightarrow \frac{\sin \theta_a}{\sin \theta_g} = \text{constant}$$

This constant is a property of the glass and is called the **refractive index** of the glass. The symbol for refractive index is n (sometimes the Greek letter μ, pronounced 'mu' is used).

The expression below defines the refractive index n for light passing from a vacuum into a substance.

$$n = \frac{\sin \theta_v}{\sin \theta_s}$$

where θ_v = angle between the ray and the normal in a vacuum

θ_s = angle between the ray and the normal in the substance

Substance	Refractive index
ice	1.31
water	1.33
crown glass	1.51–1.65
flint glass	1.53–1.93
perspex	1.50
diamond	2.42

Table 3

For practical purposes, the difference between the path change for rays passing from air and those passing from a vacuum is so small that we can ignore it: the values given for the refractive index by the two equations are effectively the same.

The refractive indices for some different substances are given in Table 3. The refractive index of a substance can be regarded as a measure of the ability of that substance to bend light: substances having a higher refractive index are those which bend the light more.

Refraction and colour

If we allow narrow beams of light of different colours to pass through a prism, we observe that red light is bent less than blue light. This shows that the refractive index n_R for red light is less than the refractive index n_B for blue light. If a beam of white light is passed through a prism, the white light is split up into a spectrum, Figure 9.20. The white light contains the range of colours red, orange, yellow, green, blue, indigo and violet. Different colours have different wavelengths. Those colours with longer wavelengths (the red end of the spectrum) are refracted less than those with shorter wavelengths. This indicates that the refractive indices are smaller for light of longer wavelengths.

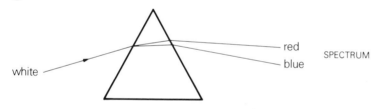

Figure 9.20 Spectrum from white light

Angles of incidence and refraction

For any ray passing from air into a substance, θ_a is the angle of incidence i, and θ_s is the angle of refraction r, Figure 9.21.

Thus, $n = \dfrac{\sin i}{\sin r}$

where i = angle of incidence

r = angle of refraction for a ray entering the substance from air (or more accurately from a vacuum)

n = refractive index of the substance.

For a ray passing from air into another substance

$$i > r$$

$$\Rightarrow \sin i > \sin r$$

$$\Rightarrow \frac{\sin i}{\sin r} > 1$$

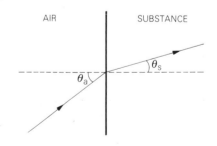

Figure 9.21

Thus the refractive index of any substance must be greater than one.

Changes in wavelength and speed on refraction

We cannot observe light waves, but interference patterns can be produced by light and this is evidence of the wave nature of light. While the wavelength of light is far too small to be observed directly, we can consider what the change in direction of the waves on refraction means in terms of wavelength. Light consists of wavefronts and the wavefronts are at right angles to the wave direction, Figure 9.22.

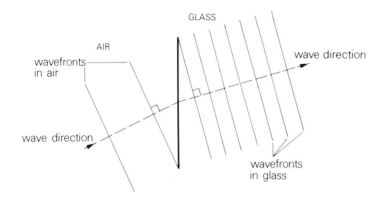

Figure 9.22 Wavefronts in air and glass

Figure 9.23 shows one wavefront AC on the air side of the boundary AB and one wavefront BD on the glass side of the boundary.

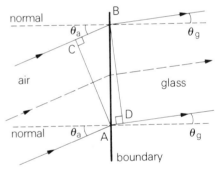

Figure 9.23

In triangle ABC, the wavefront AC is at right angles to BC and BC is one wavelength λ_a in air

$$\Rightarrow \ \sin B\hat{A}C = \frac{\lambda_a}{AB} \quad \dots [1]$$

The normal at B is at right angles to the boundary AB

$$\Rightarrow \ \theta_a = 90° - A\hat{B}C$$

But in the right-angled triangle ABC,

$$B\hat{A}C = 90° - A\hat{B}C \quad \text{(sum of internal angles } = 180°\text{)}$$

$$\Rightarrow \qquad \theta_a = B\hat{A}C$$

So, from equation [1]

$$\sin \theta_a = \frac{\lambda_a}{AB} \quad \dots [2]$$

Similarly, from the right-angled triangle ABD in which $AD = \lambda_g$

$$\sin \theta_g = \frac{\lambda_g}{AB} \quad \dots [3]$$

Dividing equation [2] by equation [3] gives the refractive index n.

$$\frac{\sin \theta_a}{\sin \theta_g} = \frac{\lambda_a}{\lambda_g} = n$$

The wavefronts in the glass are generated by wavefronts in the air, and so the number of wavefronts generated per second is the same in both the glass and the air, and the frequency f is unchanged.

$$\Rightarrow \quad \frac{\sin \theta_a}{\sin \theta_g} = \frac{f\lambda_a}{f\lambda_g} = n$$

From the wave equation $v = f\lambda$

$$v_a = f\lambda_a \quad \text{where } v_a = \text{speed of light in air}$$
$$v_g = f\lambda_g \quad\quad\quad v_g = \text{speed of light in glass}$$
$$\Rightarrow \quad \frac{\sin\theta_a}{\sin\theta_g} = \frac{v_a}{v_g} = n$$

For glass:

$$n = \frac{\text{wavelength in air (or a vacuum)}}{\text{wavelength in glass}}$$
$$= \frac{\text{speed in air (or a vacuum)}}{\text{speed in glass}}$$

This is true for any medium of refractive index n:

$$n = \frac{\text{wavelength in a vacuum}}{\text{wavelength in the medium}}$$
$$= \frac{\text{speed in a vacuum}}{\text{speed in the medium}}$$

Total internal reflection

If a ray of light will travel in one direction, it could travel along the same path in the opposite direction. The semicircular glass block can be used to investigate the bending of a ray of light at one surface only, and we can consider what happens to a ray passing from glass to air, Figure 9.24.

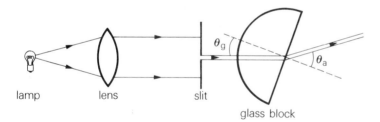

Figure 9.24 Ray passing from glass to air

If θ_g is varied and pairs of values of θ_g and θ_a are measured, a set of values such as those shown in Table 4 might be obtained.

When θ_g is increased above 36°, the ray of light does not emerge from the glass block, but is reflected at the straight face, Figure 9.25. The ray is said to be totally internally reflected.

θ_g	θ_a
10°	17°
16°	28°
23°	42°
31°	61°
35°	77°

Table 4

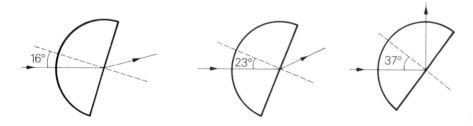

Figure 9.25 Total internal reflection

For the ray to emerge from the glass, θ_a must be less than 90°. Thus the maximum angle of incidence θ_g for the ray passing from glass to air can be calculated using the following equation.

$$n = \frac{\sin \theta_a}{\sin \theta_g}$$

For maximum value, $\theta_a = 90°$

$\Rightarrow$ $\qquad \sin \theta_a = 1$

$\Rightarrow$ $\qquad n = \dfrac{1}{\sin \theta_g}$

or $\sin \theta_g = \dfrac{1}{n}$

The maximum value of θ_g for which refraction occurs is called the critical angle C.

$$\sin C = \frac{1}{n}$$

$\Rightarrow$ $\qquad C = \sin^{-1}\left(\dfrac{1}{n}\right)$

For θ_g greater than C, the ray undergoes total internal reflection.

9.5 Intensity of illumination

When an object is illuminated, it is receiving light energy. The **intensity** of the illumination of a surface is defined as the amount of light energy per second falling on one square metre of the surface. Since energy per second is power, the intensity of illumination is a measure of the power per unit area and is measured in watts per square metre ($W\,m^{-2}$).

Inverse square law

Figure 9.26 shows an experimental arrangement that can be used to investigate the relationship between the intensity of illumination on a surface and the distance of the light source from the surface. The experiment must be carried out in a darkened room so that the lamp is the only source of light.

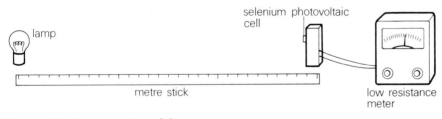

Figure 9.26 Investigation of the variation of intensity with distance

When light shines on the selenium photovoltaic cell, the cell produces an e.m.f. which is proportional to the intensity of illumination on it. The low resistance meter measures this e.m.f. and the meter reading is directly proportional to the intensity of illumination on the cell. The distance from the lamp to the cell is measured and the reading on the meter is noted. If this is repeated for a number of different distances and a graph of the results is plotted, the graph shown in Figure 9.27 is obtained.

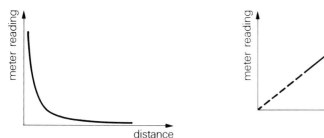

Figure 9.27 Variation of meter reading with distance between the cell and the lamp

Figure 9.28 Results of experiment demonstrating inverse square law

The relationship between the meter reading and the distance is not apparent from this graph, but an inverse relationship is indicated by the fact that the meter reading decreases as the distance increases. If a graph of the meter reading against $1/(\text{distance})^2$ is plotted, a straight line is obtained which, when extended, passes through the origin, Figure 9.28.

Since the meter reading is directly proportional to the intensity of illumination on the cell, this result shows that

$$\text{intensity} \propto \frac{1}{(\text{distance})^2}$$

This is an expression of the **Inverse Square Law**, which states that the intensity of the illumination is inversely proportional to the square of the distance from the light source.

The Inverse Square Law applies only to light radiating out from a point source, Figure 9.29.

In the experiment to show the Inverse Square Law, the distances measured are large compared with the size of the lamp filament from which the light is emitted: the lamp is effectively a point source. In a parallel beam of light the energy is not spreading out and the intensity does not decrease with distance from the source, Figure 9.30.

Figure 9.29 Light radiated from a point source

Figure 9.30 A parallel beam of light

Summary

When an object is viewed in a plane mirror, the perpendicular distance from the image to the mirror equals the perpendicular distance from the object to the mirror.

A real image is an image formed at a position from which the rays of light actually come.

A virtual image is an image formed at a position from which the rays appear to come rather than at a position from which the rays actually come.

The Laws of Reflection state that:
1 the incident ray, the reflected ray and the normal at the point of reflection are all in the same plane;
2 the angle of incidence equals the angle of reflection.

For a ray of light crossing the boundary between air (or vacuum) and a substance, the refractive index n of the substance is given by

$$n = \frac{\sin \theta_a}{\sin \theta_b}$$
where θ_a = angle between the ray in air and the normal to the boundary,

θ_b = angle between the ray in the substance and the normal to the boundary.

The refractive index of a substance is greater for light of higher frequency. For a ray of light passing the boundary between air (or vacuum) and a substance of refractive index n

$$n = \frac{\text{wavelength in air (or vacuum)}}{\text{wavelength in the substance}}$$

$$n = \frac{\text{speed in air (or vacuum)}}{\text{speed in the substance}}$$

Within a transparent substance, a ray undergoes total internal reflection if the angle between the ray and the normal is greater than the critical angle C given by

$$C = \sin^{-1}\left(\frac{1}{n}\right)$$

The intensity of illumination of a surface is the amount of light energy falling on one square metre of the surface in one second, and is measured in watts per square metre (W m^{-2}).

The Inverse Square Law states that the illumination produced on a surface by a point light source is inversely proportional to the square of the distance between the surface and the light source.

Problems

1 The diagram represents a ray of light passing from a liquid into air. Calculate the refractive index of the liquid.

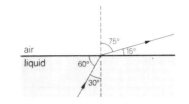

2 If the speed of light in air is $3.0 \times 10^8 \text{ m s}^{-1}$, what is the speed of light in glass of refractive index 1.5?

3 State what happens to the speed, wavelength and frequency of light when it passes from air to glass.

4 When a beam of light crosses the boundary from glass to air, what happens to its direction if it is not travelling along the normal?

5 A beam of red light of wavelength $6.25 \times 10^{-7} \text{ m}$ in air enters a glass block. The speed of light in air is $3.0 \times 10^8 \text{ m s}^{-1}$ and the refractive index of the glass is 1.5.
What is:
a) the frequency of the red light in air;
b) the frequency of the red light in glass;
c) the wavelength of the red light in the glass;
d) the speed of the red light in the glass?

6 Copy and complete the following diagrams to show the paths of rays of light when they cross both of the boundaries between the substance and air. Calculate and mark in the angles. (Refractive index of glass = 1.60; refractive index of perspex = 1.50)

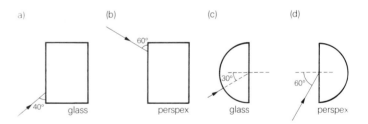

7 a) The refractive index for water is 1.33. Calculate the critical angle for water.
b) Copy and complete the following diagrams to show the paths of rays after they have met the boundary between the water and air. Mark in the values of the angles between the rays and the normals.

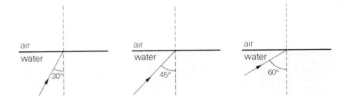

8 A ray of light from a tungsten filament lamp is incident on a glass prism as shown in the diagram.

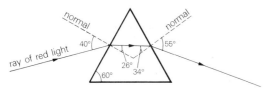

The refractive index of glass is 1.53 for blue light and 1.51 for red light.

a) If P and Q represent the ends of the visible spectrum, which is the blue end?

b) Calculate the angle α

c) From the refractive indices above, deduce whether red or blue light travels faster through the glass. Show your reasoning.   *SCEEB*

9 The diagram shows the path of a ray of red light passing through a glass prism.

a) Use the information in the diagram to find the refractive index of the glass for the red light.

b) The refractive index of the glass for blue light is 1.58. Draw a diagram similar to the one above to show the path of a ray of blue light through the glass prism. Make the initial angle of incidence 40° as before.   *SCEEB*

10 The diagram shows the ray AOB traced by a pupil investigating the refraction of red light using a semicircular glass block.

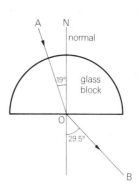

a) Use the information given in the diagram to calculate the refractive index of glass, $_{air}n_{glass}$, for red light.

b) Draw an accurate diagram to show the path of the ray if angle AON is increased to 30°.

c) The speed of red light in air is $3 \times 10^8 \, \text{m s}^{-1}$. Calculate the speed of red light in glass.   *SCEEB*

11 The intensity of illumination at a distance of 1 m from a lamp is $2 \, \text{W m}^{-2}$. What would be the intensity of illumination at a distance of 4 m from the lamp?

12 If a lamp produces an intensity of illumination of $16 \, \text{W m}^{-2}$ at a distance of 2 m from the lamp, calculate the intensity at distances of 1 m, 3 m, 4 m and 5 m.
Use these results to plot a graph to show how the intensity varies with distance from the lamp.

10 Lenses

10.1 Types of lenses

Lenses come in a variety of shapes and sizes, Figure 10.1.

Figure 10.1 Some different lenses

The lenses can be divided into two basic groups; converging lenses and diverging lenses. Lenses which are thicker at the centre than at the edges are converging lenses, and those that are thinner at the centre are diverging. In Figure 10.1, the first three lenses are converging and the last three are diverging.

Principal axis

In considering the effects of lenses on rays of light, it is useful to take, as a reference direction, a straight line through the centre of the lens at right angles to the surface of the lens. This line is called the **principal axis** of the lens. Figure 10.2 shows the principal axes of two lenses.

Figure 10.2 Principal axes

10.2 Effects of converging lenses on light

Rays which are parallel to the principal axis of a converging lens are refracted by the lens in such a way that they converge to a point on the principal axis. This point is called the **principal focus** of the lens, and the distance from the centre of the lens to the principal focus is called the **focal length** f, Figure 10.3.

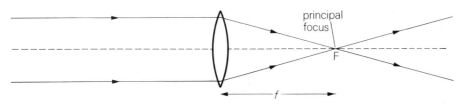

Figure 10.3 Principal focus and focal length

If rays parallel to the principal axis are directed to the lens, they converge to a point at the principal focus. Any lens has two principal foci, one on either side of the lens and each at the same distance from the centre of the lens.

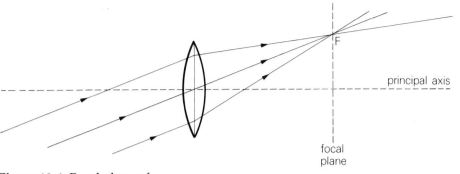

Figure 10.4 Focal plane of a converging lens

Figure 10.5 Centre of lens

A converging lens focuses parallel rays of light to a point in the focal plane of the lens. The focal plane is a plane which passes through the principal focus and is perpendicular to the principal axis, Figure 10.4.

Thus, in tracing rays of light through converging lenses, we know that rays parallel to the principal axis are refracted so that they pass through the principal focus and that sets of parallel rays meet in the focal plane of the lens.

The faces of a lens at its centre are very nearly parallel if we consider only a small area, Figure 10.5. In considering how light passing through the centre of a lens will behave, it is useful to consider how light is affected by passing through a parallel-sided glass block.

When a ray passes through a parallel-sided glass block, it emerges parallel to the incident ray, Figure 10.6.

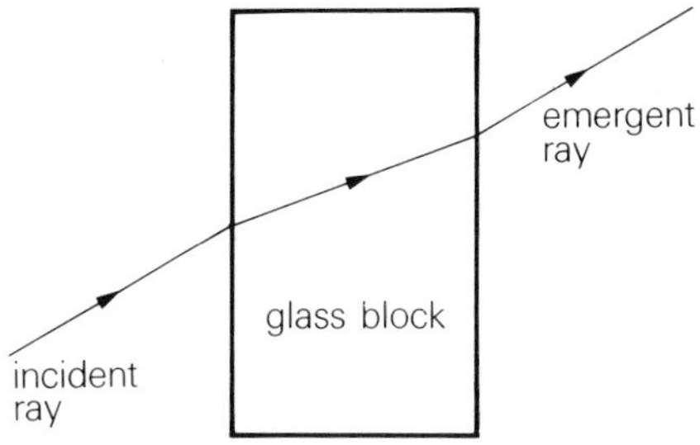

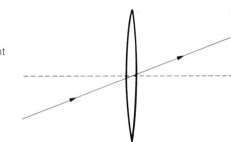

Figure 10.6 Ray through a parallel-sided glass block

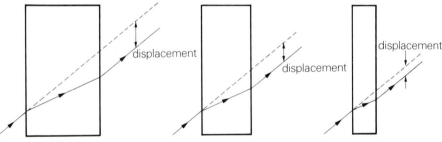

Figure 10.7 Displacements of rays by parallel-sided glass blocks

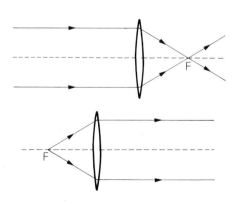

Figure 10.8 Ray through the centre of a thin lens

While the incident and emergent rays are parallel, the emergent ray is displaced sideways. The thinner the glass block, the smaller is the displacement, Figure 10.7.

The centre of a lens has two faces which are effectively parallel. If the lens is thin, the displacement of the ray is small. Thus, for a thin lens, a ray through the centre does not change direction, and its displacement is negligible, Figure 10.8.

The paths of rays passing through the centre of the lens and those which are parallel to the principal axis of the lens are particularly useful in determining where images are formed by lenses.

Since rays parallel to the principal axis are refracted so that they pass through the principal focus, the converse is also true. Rays from the principal focus of a lens will emerge, from the lens, parallel to the principal axis, Figure 10.9.

When considering image formation by lenses, a convenient object to use is

Figure 10.9 Rays parallel to the principal axis

an arrow because it is simple, has definite size (length) and can be seen to be upright or inverted (upside down). Figure 10.10 shows two rays from a point at the tip of an arrow OA passing through a lens.

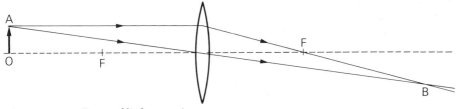

Figure 10.10 Rays of light passing through a lens from an object A

The two rays plotted are those for which we know the paths, that is the ray parallel to the principal axis and that through the centre of the lens. They are only two of the many rays from the same point which pass through the lens and meet at B, Figure 10.11.

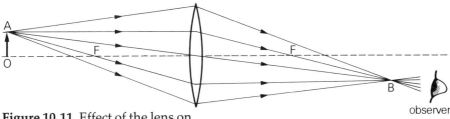

Figure 10.11 Effect of the lens on rays from the object

For the observer on the right, the rays come from point B, although they originally came from point A.

Consider the two rays shown in Figure 10.12.

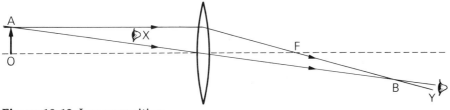

Figure 10.12 Image position

The observer at X sees them as coming from the point A. If the observer is at Y, he will see them as coming from B. Thus, when the observer views the object through the lens, he sees an image of the point A at B. Tracing rays from other parts of the object shows how an image of the object is made up at I, Figure 10.13.

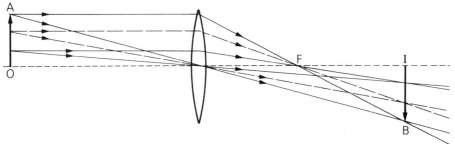

Figure 10.13 Formation of an image

An image can be located by tracing two rays from the tip of the object, one parallel to the principal axis and the other through the centre of the lens. Where these two rays meet, after passing through the lens, is the location of the tip of the image. If the object is perpendicular to the principal axis, and has its base on the principal axis, the image will also be perpendicular to the principal axis and have its base on the principal axis. Thus, if we locate the top of the image, we effectively locate the whole image, and can see whether the image is upright or inverted, magnified or diminished.

In the example shown in Figure 10.13, the observer sees the tip of the image as being located at B and the rays of light actually come from B. This type of image is called a **real image**. When the image is formed at a position from which the rays of light actually come, the image is called a real image.

Virtual image formation by a converging lens

Figure 10.14 shows a ray diagram for an object between the focal plane and the converging lens. That is to say, the object distance is less than the focal length.

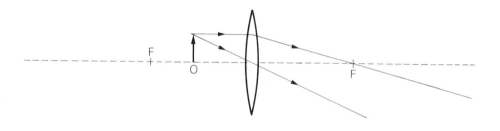

Figure 10.14 Object distance less than the focal length

The lens has had a converging effect on the rays, but the converging effect is not sufficient to make the rays meet after passing through the lens. Again we locate the image by considering from where, to the observer, the rays appear to come. This point is located by extending the rays back as shown in Figure 10.15.

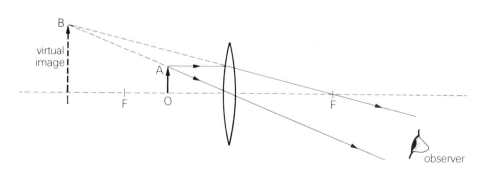

Figure 10.15

The rays from A appear to the observer, to come from B, and thus B is the image of A. The observer sees the image IB of object OA. Since the rays only appear to come from the image and do not actually do so, the image IB is a virtual image.

Examples of images formed by converging lenses

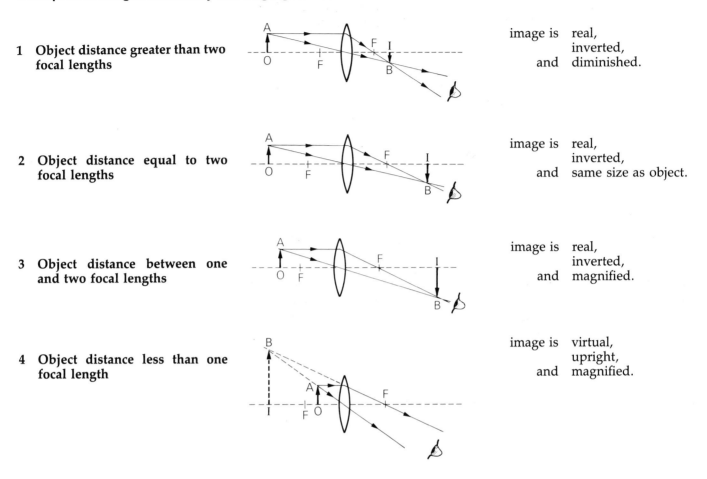

1 **Object distance greater than two focal lengths**

image is real, inverted, and diminished.

2 **Object distance equal to two focal lengths**

image is real, inverted, and same size as object.

3 **Object distance between one and two focal lengths**

image is real, inverted, and magnified.

4 **Object distance less than one focal length**

image is virtual, upright, and magnified.

10.3 Effects of diverging lenses on light

Rays of light that are parallel to the principal axis of a diverging lens are not brought to a focus, but the rays which have passed through the lens appear to an observer to diverge from a single point, Figure 10.16. This point is called the **principal focus** of the lens and is denoted by the symbol F.

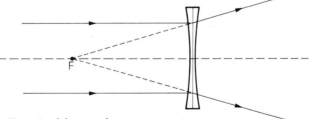

Figure 10.16 Principal focus of a diverging lens

As for a converging lens, each diverging lens has two principal foci at equal distances from the centre of the lens; the distance from the focus to the centre of the lens is called the focal length. With a diverging lens, the principal focus is said to be 'virtual' or 'imaginary', because rays of light parallel to the principal axis only appear to come from the principal focus.

Parallel rays which pass through a diverging lens appear to come from a point on the focal plane of the lens. The focal plane is a plane through the principal focus and perpendicular to the principal axis, Figure 10.17.

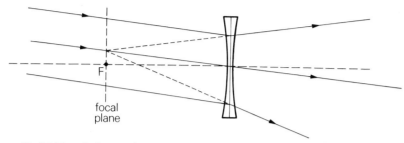
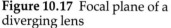

Figure 10.17 Focal plane of a diverging lens

As with a converging lens, rays of light through the centre of a thin diverging lens are unaffected.

Using these facts, we can produce ray diagrams for diverging lenses. It is important to remember that the location of a point on the image is the point from which, to the observer, the rays appear to be coming. Thus, if the rays do not actually meet, the image is found by extending the rays back until they do meet.

Example 1

An object is viewed through a diverging lens of focal length 10 cm; the object distance is 12 cm. Draw an accurate ray diagram to find the image distance.

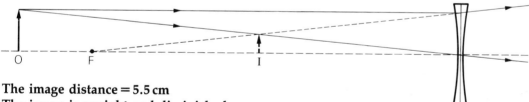

The image distance = 5.5 cm
The image is upright and diminished.

The image location is where the rays appear to come from rather than where they actually come from. Thus the image is **virtual**.

Example 2

An object is viewed through a diverging lens of focal length 12 cm; the object distance is 9 cm. Draw an accurate ray diagram to find the image distance.

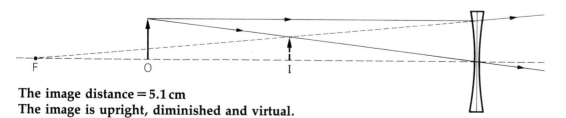

The image distance = 5.1 cm
The image is upright, diminished and virtual.

In both of these examples, one for an object distance greater than the focal length and one for an object distance less than one focal length, the image obtained is virtual. In fact, for any real object, the image formed by a diverging lens is virtual.

10.4   Images of objects at great distances

Figure 10.18 shows rays of light from points at different distances entering the eye of an observer.

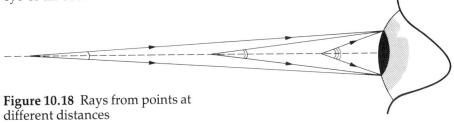

Figure 10.18 Rays from points at different distances

From this it is apparent that, as the object distance increases, the angle between the rays decreases. For long object distances, the angles between the rays are so small that the rays can effectively be considered as being parallel.

Parallel rays are said to come from infinity or to meet at infinity. Thus, for an object at infinity, a lens forms an image in the focal plane, Figure 10.19.

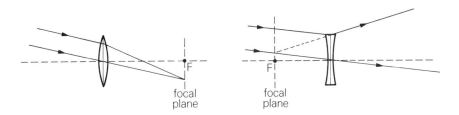

Figure 10.19 Rays from an object at infinity

The reverse is also true for a converging lens but not for a diverging lens. An object in the focal plane of a converging lens produces an image at infinity (Figure 10.20). An object in the focal plane of a diverging lens produces a virtual upright image half-way between the focal plane and the centre of the lens; the image is half the size of the object, Figure 10.21.

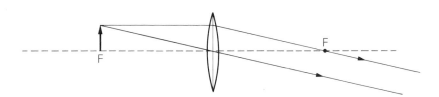

Figure 10.20 Rays from an object in the focal plane of a converging lens

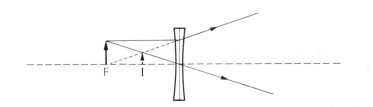

Figure 10.21 Rays from an object in the focal plane of a diverging lens

10.5   The lens equation

The lens equation relates the object distance u, the image distance v and the focal length f. If one of these quantities is unknown, its value can be found from the other two either by drawing a ray diagram, or by using the lens equation.

Figure 10.22 shows a ray diagram for a converging lens in which an object OA lies between one and two focal lengths from the lens.

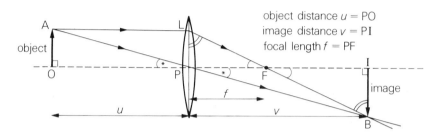

object distance u = PO
image distance v = PI
focal length f = PF

Figure 10.22

Triangles LPF and IBF are similar

$$\Rightarrow \qquad \frac{IB}{LP} = \frac{IF}{PF} = \frac{v-f}{f}$$

But LP = OA = object size, and IB = image size

$$\Rightarrow \qquad \frac{\text{image size}}{\text{object size}} = \frac{v-f}{f} \qquad \ldots[1]$$

Triangles OAP and IBP are similar

$$\Rightarrow \qquad \frac{\text{image size}}{\text{object size}} = \frac{IB}{OA} = \frac{v}{u} \qquad \ldots[2]$$

From equations [1] and [2]

$$\frac{v}{u} = \frac{v-f}{f}$$

$$\Rightarrow \qquad \frac{v}{u} = \frac{v}{f} - \frac{f}{f}$$

$$\Rightarrow \qquad \frac{v}{u} = \frac{v}{f} - 1$$

$$\Rightarrow \qquad \frac{v}{u} + 1 = \frac{v}{f}$$

Divide both sides of the equation by v

$$\frac{1}{u} + \frac{1}{v} = \frac{1}{f} \qquad \ldots[3]$$

Equation [3] is the **lens equation**. It has been derived for this one particular situation, but it applies to **any** lens situation if the following conventions are applied.

f is positive for a converging lens, but negative for a diverging lens,

u is positive for a real object, but negative for a virtual object,

v is positive for a real image, but negative for a virtual image.

Example 3

An object is placed 24 cm from a converging lens of focal length 8 cm. Find from the lens equation the distance of the image from the lens. Is the image real or virtual?

$u = 24$ cm, $f = 8$ cm, $v = ?$

$$\frac{1}{u} + \frac{1}{v} = \frac{1}{f}$$

$$\Rightarrow \frac{1}{24} + \frac{1}{v} = \frac{1}{8}$$

$$\Rightarrow \frac{1}{v} = \frac{1}{8} - \frac{1}{24} = \frac{3-1}{24} = \frac{2}{24}$$

$$\Rightarrow v = \frac{24}{2} = 12$$

The distance of the image from the lens is 12 cm
The image is **real** because v is positive.

Example 4

A real object is viewed through a diverging lens of focal length 12 cm. Find the image distance from the lens equation if the object distance is 36 cm. Is the image real or virtual?

$u = 36$ cm, $f = -12$ cm, $v = ?$

$$\frac{1}{u} + \frac{1}{v} = \frac{1}{f}$$

$$\Rightarrow \frac{1}{36} + \frac{1}{v} = \frac{1}{-12}$$

$$\Rightarrow \frac{1}{v} = -\frac{1}{12} - \frac{1}{36} = \frac{-3-1}{36} = \frac{-4}{36}$$

$$\Rightarrow v = -\frac{36}{4} = -9$$

Since v is negative, the image is virtual.

A virtual image is formed 9 cm from the lens

Example 5

When an object is placed 4 cm from a lens, a virtual image is formed at a distance of 5 cm from the lens. Use the lens equation to find the focal length of the lens and whether the lens is converging or diverging.

$u = 4$, $v = -5$, $f = ?$

$$\frac{1}{u} + \frac{1}{v} = \frac{1}{f}$$

$$\Rightarrow \frac{1}{4} + \frac{1}{-5} = \frac{1}{f}$$

$$\Rightarrow \frac{1}{f} = \frac{1}{4} - \frac{1}{5} = \frac{5-4}{20} = \frac{1}{20}$$

$$\Rightarrow f = 20$$

Since f is positive, the lens is converging.

The lens is converging and of focal length 20 cm

10.6 Magnification

The magnification m produced by a lens is defined as the ratio of the image size to the object size.

$$\text{magnification} = \frac{\text{image size}}{\text{object size}}$$

From equation [2] on page 119

$$m = \frac{\text{image size}}{\text{object size}} = \frac{v}{u}$$

If a combination of two lenses is used, the image formed by the first lens acts as an object for the second lens.

Thus for the first lens with an object of size x

first image size $= m_1 x$

and for the second lens

final image size $= m_1 x \times m_2$

where m_1 and m_2 are the magnification of the first lens and the second lens respectively.

For the combination of the two lenses

$$\text{total magnification} = \frac{\text{final image size}}{\text{original object size}}$$

$$= \frac{m_1 m_2 x}{x}$$

$$= m_1 m_2$$

In general for any system of lenses, the total magnification is equal to the product of the magnifications produced by individual lenses.

Example 6

An object of height 4 cm is viewed through a converging lens of focal length 5 cm held at a distance of 15 cm from the object. Find the height of the image formed.

$u = 15\,\text{cm},\ f = 5\,\text{cm},\ v = ?$

$$\frac{1}{u} + \frac{1}{v} = \frac{1}{f}$$

$$\Rightarrow \quad \frac{1}{15} + \frac{1}{v} = \frac{1}{5}$$

$$\Rightarrow \quad \frac{1}{v} = \frac{1}{5} - \frac{1}{15} = \frac{3-1}{15} = \frac{2}{15}$$

$$\Rightarrow \quad v = \frac{15}{2} = 7.5$$

$$m = \frac{\text{image size}}{\text{object size}} = \frac{v}{u} = \frac{7.5}{15} = \frac{1}{2}$$

But the object size is 4 cm

$\Rightarrow$ image size $= \frac{1}{2} \times 4 = 2$

The height of the image is 2 cm

10.7 The power of a lens

The power of a lens is a measure of how much the rays are caused to be bent by the lens. Figure 10.23 shows lenses of increasing power and the effect they have on parallel rays of light.

The more powerful a lens is, the shorter is its focal length.

The power of a lens is defined as the inverse of its focal length. If the focal length is in metres, the power of the lens is in dioptres (D)

$$\text{power (D)} = \frac{1}{\text{focal length (m)}}$$

The same sign convention is used for the power of a lens as that adopted for the lens equation. The power of a converging lens is positive; the power of a diverging lens is negative.

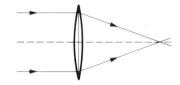

Example 7

A real image is formed when an object is viewed through a certain lens. Find the power of the lens if the object distance is 24 cm and the image distance is 12 cm.

$$u = 24\,\text{cm}, \ v = 12\,\text{cm}, \ f = ?$$

$$\frac{1}{u} + \frac{1}{v} = \frac{1}{f}$$

$$\Rightarrow \frac{1}{24} + \frac{1}{12} = \frac{1}{f}$$

$$\Rightarrow \frac{1}{f} = \frac{1+2}{24} = \frac{3}{24} = \frac{1}{8}$$

$$\Rightarrow f = 8\,\text{cm} = 0.08\,\text{m}$$

$$\text{power (D)} = \frac{1}{\text{focal length (m)}}$$

$$= \frac{1}{0.08} = 12.5$$

The power of the lens is +12.5 D

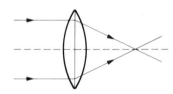

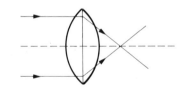

Figure 10.23 Lenses of different power

Summary

A converging lens is thicker at the centre than at the edge. A diverging lens is thinner at the centre than at the edge.

The principal axis is a straight line through the centre of a lens and at right angles to its surface.

For a converging lens, rays parallel to the principal axis converge to a point on the principal axis called the principal focus.

For a diverging lens, rays parallel to the principal axis diverge and appear to come from a point on the principal axis called the principal focus.

The focal plane is a plane at right angles to the principal axis and passing through the principal focus of the lens. All parallel rays passing through a converging lens are focused at a point in the focal plane. All parallel rays passing through a diverging lens appear to come from a point in the focal plane.

A ray through the centre of a thin lens does not change direction and its displacement is negligible.

An image can be located by a ray diagram showing a ray through the lens centre and a ray parallel to the principal axis. Alternatively, the lens equation can be used:

$$\frac{1}{u} + \frac{1}{v} = \frac{1}{f}$$

The focal length f is positive for a converging lens and negative for a diverging lens. The image distance v is positive for a real image and negative for a virtual image. The object distance u is positive for a real object and negative for a virtual object.

$$\text{magnification} = \frac{\text{image size}}{\text{object size}}$$

$$\text{For a lens, magnification} = \frac{\text{image distance}}{\text{object distance}}$$

The power of a lens in dioptres is the inverse of its focal length in metres

$$\text{power (D)} = \frac{1}{\text{focal length (m)}}$$

Problems

1   For each of the following lenses state whether it is converging or diverging.

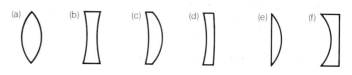

2   **a)** Draw ray diagrams to find the image distances in the following cases.
    i) an object 1 cm high at a distance of 8 cm from a converging lens of focal length 5 cm.
    ii) an object 2 cm high at a distance of 8 cm from a converging lens of focal length 5 cm.
  b) From your answers in (a), what can you conclude about the effect of the size of the object on the image distance?

3   **a)** Draw a ray diagram to find the image distances in the following cases.
    i) an object 10 cm away from a converging lens of focal length 5 cm,
    ii) an object 8 cm away from a converging lens of focal length 4 cm.
  b) From your answers in (a), what can you conclude about the relationship between the image distance and the object distance when the object distance is double the focal length?

4   What is the relationship between the object distance and the focal length of a converging lens in the following cases?
    i) the image formed is real, inverted and diminished,
    ii) the image formed is real, inverted and the same size as the object,
    iii) the image formed is real, inverted and magnified,
    iv) the image formed is virtual, upright and magnified.

5   **a)** An object is viewed through a converging lens of focal length 10 cm. Draw ray diagrams to find the image distance for the following object distances.
    i) $u = 24$ cm, **ii)** $u = 20$ cm, **iii)** $u = 15$ cm, **iv)** $u = 6$ cm
  b) Check your answers using the lens equation.

6   **a)** An object is viewed through a diverging lens of focal length 5 cm. Draw ray diagrams to find the image distance for the following object distances.
    i) $u = 12$ cm, **ii)** $u = 10$ cm, **iii)** $u = 5$ cm, **iv)** $u = 4$ cm.
  b) Check your answers using the lens equation.

7   Copy and complete the following table.

type of lens	object distance	image distance	focal length	type of image	magnification
converging	15 cm	...	20 cm	...	...
...	20 cm	20 cm	...	real	...
...	10 cm	...	− 12 cm	...	...
converging	5 cm	...	...	...	2

8   Calculate the power of the following converging lenses.
  a) A lens of focal length 20 cm.
  b) A lens which produces a real image at a distance of 40 cm when the object is 5.0 cm from the lens.
  c) A lens which produces a virtual image at a distance of 30 cm from the lens when the object is 6.0 cm from the lens.

9 a) A magnified image of a colour transparency can be produced by a slide viewer, a diagram of which is shown.

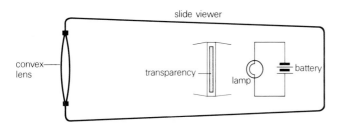

In this viewer the transparency is positioned 10 cm from the convex lens which has a focal length of 12 cm.

   i) Draw a diagram to scale illustrating how the convex lens produces a magnified image of the transparency.

   ii) Calculate the linear magnification of this image.

b) A convex lens is used in a slide projector to produce a magnified image of the transparency on a screen.

   i) State any differences between the image produced by the slide projector and that produced by the slide viewer.

   ii) For a convex lens of focal length 10 cm state the range of distances from transparency to lens which allow a magnified image to be produced on the screen.

   iii) If the sharp image produced by the projector does not fill the screen, describe the adjustments which should be made to produce a sharp image which does fill the screen. *SCEEB*

10 A box camera has a converging lens of focal length 0.14 m which produces images on a photographic film at the back of the box.

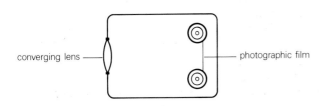

a) What is the distance between the lens and film, if a sharp image of an object 2.0 m in front of the lens is formed on the film?

b) How far must the lens be moved to form a sharp image on the film of a very distant object? *SCEEB*

11 Applications of lenses

11.1 The eye

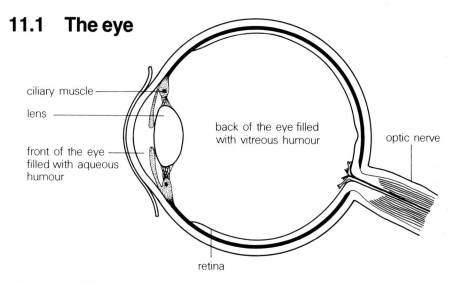

Figure 11.1 The eye

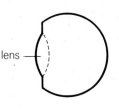

Figure 11.2 Simple model of the eye

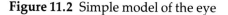

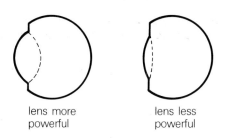

lens more powerful   lens less powerful

Figure 11.3 Changing the power of the eye lens

The converging lens at the front of the eye focuses light onto the retina which then reacts to the light by producing nerve pulses which are transmitted along the optic nerve to the brain. In fact, the vitreous humour and aqueous humour, which are clear jellies, combine with the lens to form a complex converging lens, but it is easier to understand the operation of the eye by representing the focusing parts of the eye by a simple convex lens at the front of the eye, Figure 11.2.

The shape of the lens is controlled by the ciliary muscles. These muscles are wrapped round the lens in such a way that, when they contract, they squeeze the lens and cause it to curve more, thus increasing the power of the lens. When the muscles relax, the lens takes on its original shape which is less curved and the lens power is less, Figure 11.3.

This ability to change the curvature of the lens means that the eye is able to focus an object at different distances. You will appreciate this happening if you look through a nearby window. The eye is normally focused on the scene outside, but, if your attention is drawn to a mark or some dirt on the window, your eye will focus on this and you will be less aware of the scene outside.

A normal eye is able to focus over a range of distances from infinity (very distant objects) down to a 'near point' at approximately 0.25 m. If an object is in focus, light from a point on the object will be focused to a point on the retina, Figure 11.4.

Figure 11.4 Image formation on the retina

The image formed on the retina is inverted, which means that the eye sees everything as being upside down; the brain learns to interpret the nerve pulses as a picture the correct way up.

Since the distance from the lens to the retina is constant, for any object in focus the image distance is constant and equal to the distance from the lens to the retina.

For an object at infinity the lens must focus parallel rays to a point on the retina, Figure 11.5.

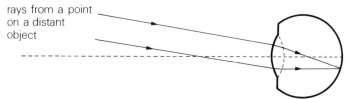

rays from a point
on a distant
object

Figure 11.5 Focused on a point at infinity

The lens equation can be applied to find how the image distance is related to the focal length of the lens.

$$\frac{1}{v} + \frac{1}{u} = \frac{1}{f}$$

for a distant object $u = \infty$

$$\Rightarrow \qquad \frac{1}{v} + \frac{1}{\infty} = \frac{1}{f}$$

$$\Rightarrow \qquad \frac{1}{v} + 0 = \frac{1}{f}$$

$$\Rightarrow \qquad \frac{1}{v} = \frac{1}{f}$$

$$\Rightarrow \qquad v = f$$

the image distance v = distance from the lens to the retina.

focal length of the lens = distance from the lens to the retina.

When the eye is focused on a distant object, the focal length of the lens is equal to the distance from the lens to the retina.

for a nearer object $u < \infty$

$$\Rightarrow \qquad \frac{1}{u} \text{ is not equal to zero}$$

$$\frac{1}{v} + \frac{1}{u} = \frac{1}{f}$$

$$\Rightarrow \qquad \frac{1}{v} = \frac{1}{f} - \frac{1}{u}$$

$$\Rightarrow \qquad \frac{1}{v} < \frac{1}{f} \quad (u \text{ is positive})$$

$$\Rightarrow \qquad v > f$$

Thus the focal length is less than the image distance.

Image distance = distance from the lens to the retina

The focal length is less than the distance from the lens to the retina.

When the eye is focused on a near object, the focal length of the lens is less than the distance from the lens to the retina.

Example 1

Find the focal length of an eye lens when the eye is focused on an object at a distance of 50.0 cm if the distance from the lens to the retina is 2.50 cm.

Distance from the lens to the retina = image distance

$\Rightarrow \qquad v = 2.5\,\text{cm} \text{ and } u = 50.0\,\text{cm}$

$$\frac{1}{v} + \frac{1}{u} = \frac{1}{f}$$

$\Rightarrow \qquad \dfrac{1}{2.5} + \dfrac{1}{50} = \dfrac{1}{f}$

$\Rightarrow \qquad \dfrac{20 + 1}{50} = \dfrac{1}{f}$

$\Rightarrow \qquad \dfrac{21}{50} = \dfrac{1}{f}$

$\Rightarrow \qquad f = \dfrac{50}{21}$

$\Rightarrow \qquad f = 2.40$

The focal length of the lens is 2.4 cm

Example 2

Find the focal length of an eye lens when the eye is focused on an object at a distance of 20.0 m if the distance from the lens to the retina is 2.50 cm.

Distance from the lens to the retina = image distance

$\Rightarrow \qquad v = 2.5\,\text{cm}$

$\Rightarrow \qquad v = 0.025\,\text{m}$

$\qquad u = 20\,\text{m}$

$$\frac{1}{v} + \frac{1}{u} = \frac{1}{f}$$

$\Rightarrow \dfrac{1}{0.025} + \dfrac{1}{20} = \dfrac{1}{f}$

$\Rightarrow \qquad \dfrac{800 + 1}{20} = \dfrac{1}{f}$

$\Rightarrow \qquad \dfrac{801}{20} = \dfrac{1}{f}$

$\Rightarrow \qquad f = \dfrac{20}{801}$

$\Rightarrow \qquad f = 0.0250$

The focal length of the lens is 2.5 cm

These two examples illustrate the fact that the focal length of a lens is less when the eye focuses on nearer objects than when it focuses on distant objects. Figure 11.6 shows rays from points on the principal axis being focused. In the case of the nearer object, the lens is more powerful since it must cause greater bending of the rays.

For a lens

$$\text{power} = \frac{1}{\text{focal length}}$$
$$\text{(D)} \qquad (\text{m}^{-1})$$

Thus the more powerful lens is that with the shorter focal length.

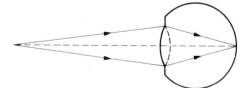

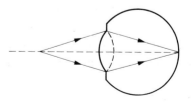

Figure 11.6 Focusing rays from objects at different distances

Example 3

Using the figures given in the two previous examples, calculate the powers of the lens for the object distances 50.0 cm and 20.0 m.

When the object distance was 50.0 cm

$$f = 0.024\,\text{m} \qquad \text{power} = \frac{1}{0.024}\text{D} \qquad \text{power} = 41.7\,\text{D}$$

For the object distance of 50.0 cm the power is +41.7 D

When the object distance was 20.0 m

$$f = 0.025\,\text{m} \qquad \text{power} = \frac{1}{0.025}\text{D} \qquad \text{power} = 40.0\,\text{D}$$

For the object distance of 20.0 m the power is +40.0 D

11.2   Short-sightedness

A person who is short-sighted is unable to focus properly on distant objects. When the eye is focused on a distant object, its lens is at its thinnest (least powerful). In the eye of a short-sighted person the lens is too powerful because its curvature cannot be sufficiently reduced and the rays from the distant object are focused in front of the retina instead of on the retina, Figure 11.7.

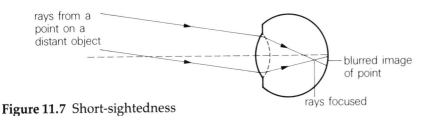

Figure 11.7 Short-sightedness

Short-sightedness can be corrected by using a diverging spectacle or contact lens. The combination of the negative power of the diverging lens and the positive power of the converging lens of the eye produces a less powerful converging effect than that of the lens of the eye alone.

The diverging lens produces an image of the distant object. The distance of this image from the eye is less than the distance of the object from the eye. This image acts as a virtual object on which the eye focuses, Figure 11.8.

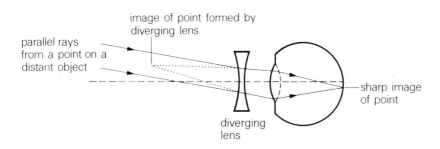

Figure 11.8 Correcting for short-sightedness

11.3 Long-sightedness

A person who is long-sighted cannot focus on nearer objects; this means that the near point is further than the normal distance of about 0.25 m from the eye. In this case the lens of the eye is not sufficiently powerful to focus rays, from near objects, onto the retina, Figure 11.9.

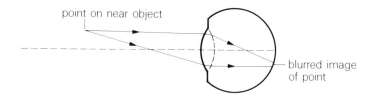

Figure 11.9 Long-sightedness

Long-sightedness can be corrected by using a converging spectacle or contact lens. The combination of this lens with the lens of the eye produces a more powerful converging effect than the lens of the eye alone.

The converging lens produces an image of the near object. The distance of this image from the eye is greater than the distance of the object from the eye. This image acts as a virtual object on which the eye focuses, Figure 11.10.

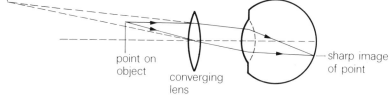

Figure 11.10 Correcting for long-sightedness

11.4 Simple camera

Like the eye, a camera has a lens system which focuses rays to form an image on a light sensitive screen. In the eye this screen is the retina and in the camera it is the film. The light produces chemical changes in the emulsion on the film and these chemical changes cause that part of the emulsion that has been exposed to light to react differently with the developer.

The lens system in a camera usually consists of a combination of several lenses, but it is easier to understand the operation of the camera if we represent the lens system by a single converging lens, Figure 11.11.

Focusing the camera

By changing its shape, the power of an eye lens is altered to focus on objects at different distances. The power of a camera lens can not be altered in this way. The camera is focused by adjusting the distance between the lens and the film. The lens mounting is on a screw thread so that its position can be

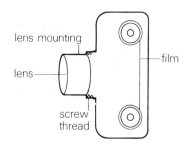

Figure 11.11 The simple camera

finely adjusted backwards and forwards. How this can be used to focus on objects at different distances can be understood by considering the lens equation.

$$\frac{1}{u} + \frac{1}{v} = \frac{1}{f} \quad \Rightarrow \quad \frac{1}{u} = \frac{1}{f} - \frac{1}{v}$$

In the camera the power and hence the focal length of the lens remains constant. Figure 11.12 shows how the image distance v changes with a change in the object distance u for a lens of fixed focal length.

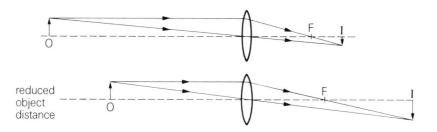

Figure 11.12 The effect of object distance on image distance

This illustrates that, for greater object distances, the image distance is less. Thus, if the camera is focused on distant objects the lens is nearer to the film than when the camera is focused on nearer objects. When focused on very distant objects the lens setting is for a distance of infinity. In this case the rays are effectively parallel and will be brought to a focus in the focal plane of the camera lens.

Using the lens equation:

$$\frac{1}{u} + \frac{1}{v} = \frac{1}{f}$$

$$u = \infty \text{ for a distant object}$$

$$\Rightarrow \frac{1}{\infty} + \frac{1}{v} = \frac{1}{f}$$

$$\Rightarrow 0 + \frac{1}{v} = \frac{1}{f} \quad \Rightarrow \quad v = f$$

Thus, when the distance setting on the camera is ∞, the distance from the lens to the film is equal to the focal length of the lens. When focused on nearer objects, the distance from the lens to the film is greater than the focal length of the lens.

Example 4

When a camera is set to focus on a distant object the distance from the centre of the lens to the film is 5.0 cm.
a) What is the power of the lens?
b) How far will the lens move when the camera is adjusted to focus on an object at a distance of 2 m?

a) When focused on a distant object $u = \infty$

$$\frac{1}{u} + \frac{1}{v} = \frac{1}{f}$$

$$\Rightarrow \frac{1}{\infty} + \frac{1}{v} = \frac{1}{f} \quad \Rightarrow \quad v = f$$

When properly focused, $v =$ distance from the lens to the film

When $u = \infty$, $v = 5.0\,\text{cm}$

$$f = 5.0\,\text{cm}$$

$$\text{power (D)} = \frac{1}{\text{focal length (m)}}$$

$$f = 0.05\,\text{m}$$

$$\text{power} = \frac{1}{0.05}$$

$$\textbf{power} = \textbf{+20 D}$$

b) $\dfrac{1}{v} + \dfrac{1}{u} = \dfrac{1}{f}$

$$f = 0.05\,\text{m}$$

$$u = 2\,\text{m}$$

$$\Rightarrow \frac{1}{v} + \frac{1}{2} = \frac{1}{0.05}$$

$$\Rightarrow \quad \frac{1}{v} = \frac{1}{0.05} - \frac{1}{2}$$

$$\Rightarrow \quad \frac{1}{v} = \frac{40 - 1}{2}$$

$$\Rightarrow \quad \frac{1}{v} = \frac{39}{2}$$

$$\Rightarrow \quad v = \frac{2}{39}$$

$$\Rightarrow \quad v = 0.051$$

$$\Rightarrow \quad v = 5.1\,\text{cm}$$

When changing the distance setting from ∞ to $2\,\text{m}$ the image distance changes from $5.0\,\text{cm}$ to $5.1\,\text{cm}$.

The lens is moved 0.1 cm further away from the film.

Exposure

The amount of light energy falling on a film depends on the exposure. In a camera this can be controlled in two ways, adjusting the 'shutter speed' or adjusting the 'aperture'.

Shutter speed

When a photograph is taken, the shutter opens for a short time, during which light enters the camera through the lens and falls on the film. By increasing the time for which the shutter is open, the light energy admitted (and hence the exposure of the film) is increased. Common exposure time settings on cameras are $\frac{1}{30}\,\text{s}$, $\frac{1}{60}\,\text{s}$, $\frac{1}{125}\,\text{s}$, $\frac{1}{250}\,\text{s}$ and $\frac{1}{500}\,\text{s}$. The time exposure is reduced to approximately half by each successive change in setting.

Aperture

The aperture of the camera is the opening through which the light passes as it travels from the lens to the film. By adjusting the size of this opening, the amount of light energy reaching the film is regulated. Increasing the aperture allows more light to pass through and results in a brighter image. The mechanism which can be adjusted to vary the size of the aperture is called a diaphragm, Figure 11.13.

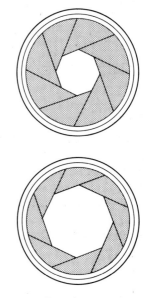

Figure 11.13 Diaphragm

Figure 11.14 illustrates how a wider aperture allows more light from a point O on an object to be focused to a point I on the film.

It is perhaps surprising to note that increasing the aperture increases the brightness but causes no increase in the size of the image produced or the scene viewed.

The different aperture settings on a camera are the different 'stop numbers'. These express the diameters of the aperture as fractions of the focal length of the lens. Some common stop numbers are f2.8, f4, f5.6, f8, f11 and f16. These represent apertures of diameters

$$\frac{1}{2.8}\times f,\ \frac{1}{4}\times f,\ \frac{1}{5.6}\times f,\ \frac{1}{8}\times f,\ \frac{1}{11}\times f \text{ and } \frac{1}{16}\times f$$

where f is the focal length of the lens.

The aperture is approximately circular in shape and its area is given by

$$\text{area} = \pi\left(\frac{d}{2}\right)^2 \text{ where } d = \text{diameter}$$

$$\text{area} = \frac{\pi}{4}\times d^2$$

$$\text{area} \propto d^2$$

Table 1 shows the squares of the diameters for successive stop numbers on a camera, with a lens system of focal length f.

From the table it can be seen that each successive increase in stop number results in halving the square of the diameter and, since

$$\text{area} \propto (\text{diameter})^2$$

the area of the aperture is halved, so that the light energy reaching the film is halved.

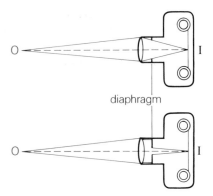

Figure 11.14 Different aperture settings

stop no.	diameter	(diameter)^2
2.8	$\frac{f}{2.8}$	$\frac{f^2}{7.8}$
4	$\frac{f}{4}$	$\frac{f^2}{16}$
5.6	$\frac{f}{5.6}$	$\frac{f^2}{31.3}$
8	$\frac{f}{8}$	$\frac{f^2}{64}$
11	$\frac{f}{11}$	$\frac{f^2}{121}$
16	$\frac{f}{16}$	$\frac{f^2}{256}$

Table 1

11.5 Astronomical telescope

A telescope is used to produce a magnified image of a distant object. When a distant object is viewed through a converging lens, the object distance u is large compared with the image distance v, Figure 11.15.

Figure 11.15 A distant object viewed through a converging lens

Magnification m is given by $m = \dfrac{v}{u}$

When u is large compared with v, the magnification is less than 1 and the image is diminished. This is illustrated in Figure 11.15. If a high magnification is to be produced, the object distance u must be small compared with the image distance. In a telescope, one lens (the objective lens) is used to produce an image of the distant object and a second lens (the eyepiece lens) is used to magnify this image.

If an object lies within the focal length of a lens, a magnified image is produced, Figure 11.16.

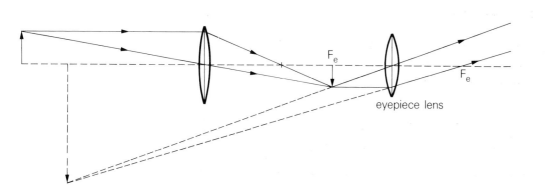

Figure 11.16 Object distance less than the focal length

The distance between the objective lens and the eyepiece lens can be adjusted so that the image formed by the objective lens falls within the focal length of the eyepiece lens, Figure 11.17. This image acts as the object for the eyepiece lens and is magnified by that lens, Figure 11.18.

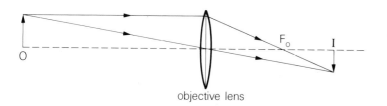

Figure 11.17 Image formed by objective lens

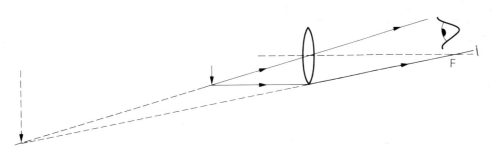

Figure 11.18 Magnification by eyepiece lens

For an astronomical telescope, the distance of the object (a star or a planet) is so great that the rays from a point on the object are effectively parallel.

Parallel rays are focused by a converging lens to form a real image in the focal plane of the lens, Figure 11.19.

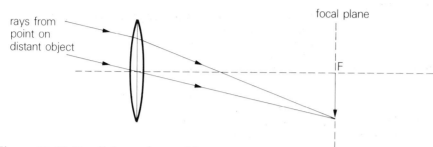

Figure 11.19 Parallel rays focused by a converging lens

In the relaxed state the eye focuses parallel rays to a point on the retina of the eye. Thus, if the telescope is adjusted so that the rays from a point on the object produce parallel rays emerging from the eyepiece, these rays will produce an image of the point on the retina.

Rays from a point in the focal plane of a converging lens will emerge from the lens as parallel rays. Thus the eyepiece is positioned so that the image formed by the objective lens lies in the focal plane of the eyepiece lens, Figure 11.20.

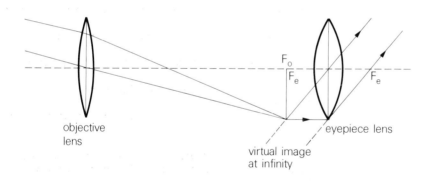

Figure 11.20 Image formation by the eyepiece lens

Thus, for an astronomical telescope focused on a very distant object, the image formed by the objective lens lies in the focal planes of both the objective and eyepiece lenses. This means that the focal planes of the two lenses coincide and **the distance between the two lenses is equal to the sum of the focal lengths of the two lenses.**

Summary

Short-sightedness is the inability of the eye to focus properly on distant objects. It can be corrected by using a diverging spectacle or contact lens.

Long-sightedness is the inability of the eye to focus properly on nearer objects. It can be corrected by using a converging spectacle or contact lens.

A simple camera has a converging lens. The camera is focused by moving the lens so that the distance from the lens to the film is the correct image distance.

The amount of light energy falling on the film in a camera is controlled by the aperture and shutter speed of the camera.

An astronomical telescope consists of two converging lenses, the objective lens and the eyepiece lens.

For an astronomical telescope focused on a distant object, the image formed by the objective lens lies in the focal planes of both the objective lens and the eyepiece lens.

Problems

1   Is the lens in the eye a converging or a diverging lens?

2   When an eye changes from focusing on a distant object to a near object, what happens to:
   a) the curvature of the eye lens,
   b) the focal length of the eye lens,
   c) the power of the eye lens?

3   Is the image formed on the retina of the eye:
   a) upright or inverted,
   b) magnified or diminished,
   c) real or virtual?

4   For each of the following cases, state whether the focal length of the eye lens is less than, equal to or greater than the distance from the eye lens to the retina.
   a) The eye is focused on a boat on the horizon.
   b) The eye is focused on a book at a distance of 25 cm.

5   In each of the following cases, state whether the person is shortsighted or longsighted, and whether a converging or diverging spectacle lens should be used to correct the defect.
   a) A girl is unable to read the number plate of a car which is on the other side of the road.
   b) A boy is unable to read the print of a newspaper which he is holding 25 cm in front of his eyes.

6   For each of the following statements, write whether it is true for an eye, a simple camera, neither or both.
   a) It focuses on objects at different distances by changing the shape of the lens.
   b) The image distance changes when it focuses on objects at different distances.
   c) The image may be either magnified or diminished.
   d) The image is always inverted.
   e) The image distance is always less than the focal length of the lens.

7   State two adjustments that can be made in a simple camera to increase the amount of light energy falling on the film when a photograph is taken.

8   An astronomical telescope has an objective lens and an eyepiece lens. To which of these two lenses do the following statements apply?
   a) It produces a near image of a distant object.
   b) It produces a magnified image.
   c) It produces a real image.
   d) Its image is inverted.

9   When an astronomical telescope is focused on a star, the distance between the eyepiece lens and the objective lens is 2 m. If the objective lens has a power of 0.6 D, what is the power of the eyepiece lens?

12 The wave nature of light

12.1 Early theories of the nature of light

The debate over what light is has continued for many centuries. We have records of this debate as far back as the ancient Greeks. Plato wrote about light as being a fire which flowed out from the eyes. Aristotle argued that this could not be the case as it did not explain why some things could be in darkness. If the eyes were the source of the light all objects could be illuminated by the eyes and nowhere would be dark.

By the start of the seventeenth century it had become generally accepted that light was received by the eye rather than transmitted by the eye. However, at that time, another debate was started about the nature of light. Scientists divided into two groups, one group believing light to be a stream of particles and the other that light was a wave. Towards the end of the seventeenth century, Huygens published his book, *Treatise on Light*. In this book he proposed the theory that light was a wave. Huygens had a formidable opponent to his wave theory. Isaac Newton supported the corpuscular theory which stated that light was a stream of particles. Each side was able to give objections to the theory of the other and these objections could not, at that time, be answered. The biggest objection to the corpuscular theory was its failure to explain how two beams of light could pass through each other without being affected. If light were streams of particles they would be expected to collide. Supporters of the wave theory could easily demonstrate, using water waves, that waves could pass through each other without being affected. Supporters of the corpuscular theory argued that light could not be a wave since it apparently did not demonstrate the property now known as diffraction. Observation of water waves showed that they were able to bend round objects, Figure 12.1. Sound, which was known to be a wave motion, also demonstrated the property of bending round corners. Such bending, however, could not be observed with light, and Newton and his supporters argued that this was proof that light was not a wave.

Figure 12.1 Diffraction of water waves

We now know that light is diffracted round the edges of obstacles but that the amount of bending is so slight that it is not easily observed. The reason for this is that the wavelength of light is very small. Figure 12.2 shows diffraction patterns for water waves of different wavelengths and demonstrates that the wave with the shorter wavelength is diffracted less.

(a) Long wavelength   (b) Short wavelength

Figure 12.2 Diffraction of waves of different wavelengths

Light waves have wavelengths of the order of 10^{-6} m, which means that they are of the order of one ten-thousandth of the wavelength of the water waves in a ripple tank. With such a small wavelength the degree of bending is very slight, and not observable by the naked eye.

Because of the status of Newton, who supported the corpuscular theory, and the apparent fact that light did not demonstrate diffraction, the wave theory of Huygens did not receive much support. It was not until the start of the nineteenth century that Thomas Young supplied important new experimental evidence which supported the wave theory, and scientists accepted the idea that light was a wave motion. Young demonstrated that light could form an interference pattern and this could only be explained in terms of the wave theory.

12.2   Interference of waves

Interference of waves can be demonstrated with water waves in a ripple tank. An interference pattern is produced by two overlapping circular wave patterns from sources of the same frequency, Figure 12.3.

Two sources of identical circular wave patterns can be produced by passing a plane wave through two narrow slits. If the width of each slit is less than the wavelength of the wave, two circular wave patterns are produced by diffraction of the wave by the slits.

Figure 12.4 shows the interference pattern with lines drawn along areas of uniform illumination. This uniform illumination means that, along these lines, the water is calm.

These are lines along which destructive interference occurs. This results from the waves being half a wavelength out of phase and cancelling each other out.

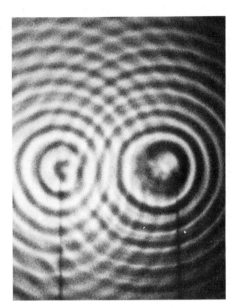

Figure 12.3 Interference of water waves

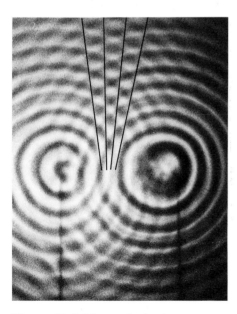

Figure 12.4 Lines of calm in an interference pattern

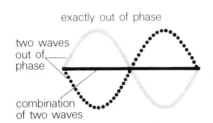
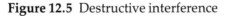

Figure 12.5 Destructive interference

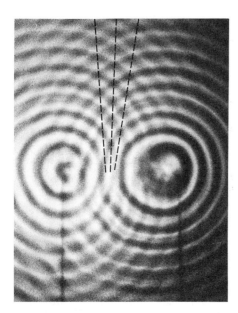

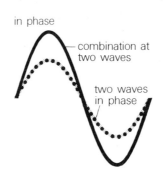

Figure 12.7 Constructive interference

Figure 12.6 Lines of crests and troughs in an interference pattern

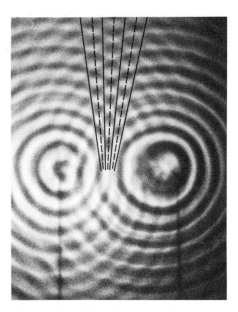

Figure 12.8 Interference pattern

Figure 12.6 shows the interference pattern with lines drawn along areas where the illumination is alternately bright and dark, showing that the water surface has a series of crests and troughs.

These are lines along which constructive interference occurs. This results from the waves being in phase and combining to give a wave of greater amplitude, Figure 12.7.

Thus the interference pattern consists of a series of lines of constructive interference and destructive interference which means that these lines are lines of maximum and minimum wave amplitude, Figure 12.8.

Young's slits experiment

Young demonstrated the interference of light by splitting a narrow beam of sunlight into two beams. He allowed sunlight to pass through a small hole made by piercing a card with a needle. Across the small hole he placed a very thin card, edge on, to divide the beam of light into two narrow beams. When he did this, Young observed a series of bright and dark lines on the wall opposite. These lines are known as interference fringes, the bright lines being produced by constructive interference and the dark lines resulting from destructive interference.

A similar experiment may be repeated in the laboratory to demonstrate the interference of light. Two narrow slits, very close together, are prepared by using a pin point to scratch two fine lines across a glass slide which has been painted black, Figure 12.9. Light is passed through these slits and is viewed on a screen, Figure 12.10.

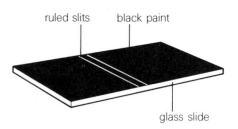

Figure 12.9 Preparation of glass slide for Young's slits experiment

Figure 12.10 Apparatus to demonstrate the Young's slits experiment

A series of fringes is observed on the screen. With a white light source, the central fringe is white but the other fringes have coloured edges. Interference patterns for different colours of light are produced by inserting different filters between the lamp and the glass slide. When this is done, it is observed that the spacing of the fringes varies with the colour of light. The fringes for light nearer the red end of the spectrum are further apart than fringes for light nearer the blue end of the spectrum. Red light has a longer wavelength than blue light and longer wavelengths produce maxima which are more widely spaced, Figure 12.11.

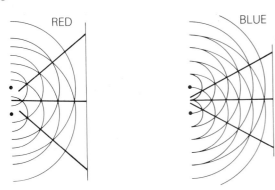

Figure 12.11 Interference patterns for waves of different wavelengths.

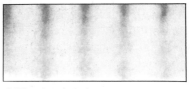

a) Slits close together

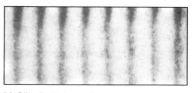

b) Slits further apart

Figure 12.12 Effect of slit separation on fringe spacing

When glass slides with slits of different separation are used, the spacing of the fringes is found to depend on the separation of the slits. The fringes are more widely spaced when the slits are closer together, Figure 12.12.

Another factor which affects the fringe separation is the distance from the slits to the screen. If the screen is moved further from the slits, the fringe separation increases.

The factors found experimentally to affect the fringe separation are listed below.
a) Red light produces fringes of greater separation than those produced by blue light:
b) Smaller slit separation produces greater fringe separation;
c) Greater distance between the screen and the slits produces greater fringe separation.

Thus the fringe separation Δx depends on the three quantities, slit separation d, the perpendicular distance from the slits to the screen D and the wavelength λ. The relationship between these quantities can be found by considering two rays from the slits S_1 and S_2 to the screen, as shown in Figure 12.13.

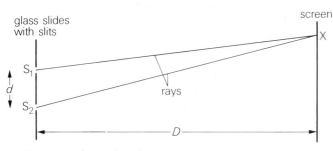

Figure 12.13 Two rays from the slits to a point on the screen

For the two rays shown in Figure 12.13, that from slit S_1 has travelled a shorter distance S_1X to the screen than the distance S_2X travelled by the ray from the slit S_2.

The difference between these two distances is called the path difference. The path difference, $S_2X - S_1X$, can be found in terms of the wavelength of the light, the slit separation and the perpendicular distance from the slits to the screen.

In Figure 12.14, the horizontal scale is different from the vertical scale in order to give a clearer picture, because D is much greater than d and x. N is the point on the screen directly opposite the point between the two slits on the glass slide; x is the distance from the point N to where the two rays meet on the screen.

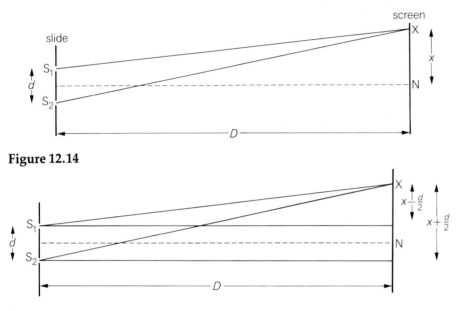

Figure 12.14

Figure 12.15

Figure 12.15 shows two triangles superimposed on Figure 12.14. Consideration of these triangles shows that:

$$(S_1X)^2 = D^2 + \left(x - \frac{d}{2}\right)^2 \qquad \text{by Pythagoras}$$

$$(S_2X)^2 = D^2 + \left(x + \frac{d}{2}\right)^2 \qquad \text{by Pythagoras}$$

$$\Rightarrow \qquad (S_2X)^2 - (S_1X)^2 = \left(x + \frac{d}{2}\right)^2 - \left(x - \frac{d}{2}\right)^2$$

$$= x^2 + xd + \frac{d^2}{4} - \left(x^2 - xd + \frac{d^2}{4}\right)$$

$$= 2xd$$

$\Rightarrow (S_2X + S_1X)(S_2X - S_1X) = 2xd$

$$\Rightarrow \qquad S_2X - S_1X = \frac{2xd}{S_2X + S_1X}$$

Since D is much greater than x, S_2X and S_1X are approximately equal to D.

$\Rightarrow S_2X + S_1X$ is approximately equal to $2D$

$\Rightarrow S_2X - S_1X = \dfrac{2xd}{2D} \quad \Rightarrow \quad S_2X - S_1X = \dfrac{xd}{D}$

This equation relates the path difference of the rays to the distance x from the centre of the interference pattern, the slit separation d and the perpendicular distance D from the slits to the screen.

If the path difference is equal to zero or a whole number of wavelengths, the waves will arrive in phase, and constructive interference will occur producing a bright fringe, Figure 12.16. Here the solid lines in the waves indicate wave fronts which left the slits at the same time.

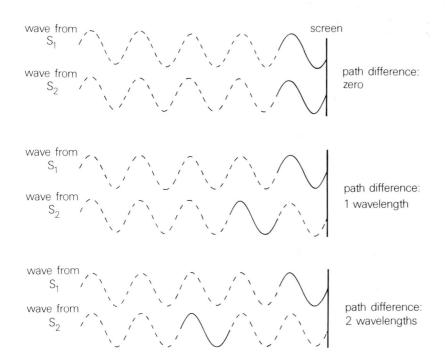

Figure 12.16 Waves from the two slits arriving in phase

Thus, for a path difference $(S_2X - S_1X)$ of zero or a whole number of wavelengths, a **bright** fringe is formed when

$$S_2X - S_1X = n\lambda \qquad \text{where } n = 0, 1, 2, 3 \ldots$$

$$\Rightarrow \qquad \frac{xd}{D} = n\lambda$$

$$\Rightarrow \qquad x = \frac{n\lambda D}{d} \quad \ldots [1]$$

The centre of the interference pattern is where $x = 0$ and $n = 0$
The first bright fringe away from the centre is that for which $n = 1$

$$x = \frac{\lambda D}{d}$$

For the next fringe, $n = 2$

$$x = \frac{2\lambda D}{d}$$

The locations of fringes are summarized in Table 1.

From the table it can be seen that, for successive fringes, the distance from the centre of the pattern increases by $\lambda D/d$. Thus the distance Δx between the

fringe	n	path difference	distance x from the centre of the pattern
central	0	0	0
1st	1	λ	$\dfrac{\lambda D}{d}$
2nd	2	2λ	$\dfrac{2\lambda D}{d}$
3rd	3	3λ	$\dfrac{3\lambda D}{d}$
$\vdots$	$\vdots$	$\vdots$	$\vdots$
nth	n	$n\lambda$	$\dfrac{n\lambda D}{d}$

Table 1 Location of bright fringes

fringes in the interference pattern produced by Young's slits is given by

$$\Delta x = \frac{\lambda D}{d}$$
   where λ = wavelength of the light

D = distance from slits to screen

d = distance between slits

White light consists of a mixture of light of different wavelengths or colours. From Table 1 it can be seen that the only fringe position that does not depend on wavelength is that for the central fringe. It is for this reason that, when using a white light source, the central fringe does not have coloured edges. For the other fringes the location depends on the wavelength, longer wavelengths producing fringes that are more widely spaced. This is why, for fringes other than the central fringe, the colours separate out and the fringes have coloured edges.

Example 1

In a Young's slits experiment, the distance from the slits to the screen is 4.0 m and the distance between the centres of the slits is 0.50 mm.
a) Find the fringe spacing for (i) violet light of wavelength 4.0×10^{-7} m; (ii) red light of wavelength 7.0×10^{-7} m.
b) Find the distance from the centre of the interference pattern to the fifth red fringe.

a) The fringe spacing is given by the equation

$$\Delta x = \frac{\lambda D}{d}$$

i) $\lambda = 4.0 \times 10^{-7}$ m for violet light

$D = 4.0$ m

$d = 0.50 \times 10^{-3}$ m

$\Rightarrow \Delta x = \frac{4 \times 10^{-7} \times 4}{0.5 \times 10^{-3}}$ $\Rightarrow \Delta x = 3.2 \times 10^{-3}$

For violet light the fringe spacing is 3.2 mm

ii) $\lambda = 7.0 \times 10^{-7}$ m for red light

$D = 4.0$ m

$d = 0.50 \times 10^{-3}$ m

$\Rightarrow \Delta x = \frac{7 \times 10^{-7} \times 4}{0.5 \times 10^{-3}}$ $\Rightarrow \Delta x = 5.6 \times 10^{-3}$

For red light the fringe spacing is 5.6 mm

b) The distance of the nth slit from the centre of the interference pattern is given by

$$x = \frac{n\lambda D}{d} \quad \text{and} \quad \Delta x = \frac{\lambda D}{d}$$

$\Rightarrow \quad x = n\Delta x$

for red light $\Delta x = 5.6 \times 10^{-3}$ m   for the fifth fringe $n = 5$

Distance of the fifth fringe from the centre of the interference pattern is given by

$\Rightarrow \quad x = 5 \times 5.6 \times 10^{-3}$ $\Rightarrow x = 28.0 \times 10^{-3}$ m

Distance from centre of interference pattern to fifth fringe is 28.0 mm.

Equations have been derived to find the distance from the centre of the interference pattern to a bright fringe and the spacing between the bright fringes. Similarly, equations can be derived to find the distance of a dark band from the centre of the interference pattern and the spacing between the dark bands.

Dark bands in an interference pattern are formed by destructive interference between the light from the two different slits S_1 and S_2. Destructive interference occurs when two waves are out of phase and this occurs when the path difference between the two rays from the slits is equal to an odd number of wavelengths.

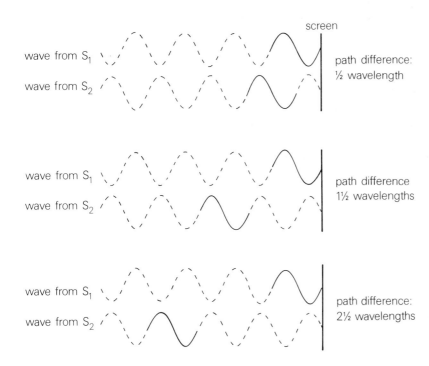

Figure 12.17 Waves from the two slits arriving out of phase

The equation for the path difference for rays from the slits S_1 and S_2 to the point X on the screen is:

$$S_2X - S_1X = \frac{xd}{D}$$

where d = distance between the slits;
x = distance of X from the centre of the interference pattern;
D = distance of the screen from the slits.

For the first dark band, this path difference is $\frac{1}{2}\lambda$

$$\Rightarrow \quad \frac{1}{2}\lambda = \frac{xd}{D} \quad \Rightarrow \quad x = \frac{\frac{1}{2}\lambda D}{d}$$

For the second dark band, the path difference is $1\frac{1}{2}\lambda$.

$$\Rightarrow \quad 1\frac{1}{2}\lambda = \frac{dx}{D} \quad \Rightarrow \quad x = \frac{1\frac{1}{2}\lambda D}{d}$$

The locations of the dark bands are summarized in Table 2

dark band	n	path difference	distance x from the centre of the pattern
1st	1	$\frac{1}{2}\lambda$	$\frac{1}{2}\frac{\lambda D}{d}$
2nd	2	$1\frac{1}{2}\lambda$	$1\frac{1}{2}\frac{\lambda D}{d}$
3rd	3	$2\frac{1}{2}\lambda$	$2\frac{1}{2}\frac{\lambda D}{d}$
nth	n	$(n-\frac{1}{2})\lambda$	$(n-\frac{1}{2})\frac{\lambda D}{d}$

Table 2 Location of dark bands

It can be seen that the distances increase by $\lambda D/d$ and the distance Δx between the **dark** bands is given by

$$\Delta x = \frac{\lambda D}{d} \qquad \dots [2]$$

Coloured fringes

White light contains the range of colours in light from violet with a wavelength of 4×10^{-7} m to red with a wavelength of 7×10^{-7} m. When Young's slits experiment is carried out with white light, multi-coloured fringes are formed. How this comes about can be illustrated by producing fringes with light of two different colours.

If Young's slits experiment is set up as follows,

distance d between the slits $= 0.1$ mm;

distance D from slits to screen $= 4$ m;

for light of wavelength λ, the fringe spacing is given by

$$\Delta x = \frac{\lambda D}{d}$$

$$\Delta x = \frac{\lambda \times 4}{0.1 \times 10^{-3}} \quad \Rightarrow \quad \Delta x = 4 \times 10^4 \times \lambda$$

Consider the fringes produced by red light of wavelength 7.0×10^{-7} m and blue light of wavelength 5.0×10^{-7} m. Table 3 gives the fringe spacing for each wavelength in metres.

Figure 12.18 is drawn to scale and shows the fringe position for the two colours. It shows that, for a mixture of red and blue light, the interference pattern consists of areas of dark, areas of red, areas of blue and areas of mixed red and blue.

	red	blue
λ	7.0×10^{-7}	5.0×10^{-7}
Δx	2.8×10^{-2}	2.0×10^{-2}
x for $n = 1$	2.8×10^{-2}	2.0×10^{-2}
x for $n = 2$	5.6×10^{-2}	4.0×10^{-2}
x for $n = 3$	8.4×10^{-2}	6.0×10^{-2}

Table 3

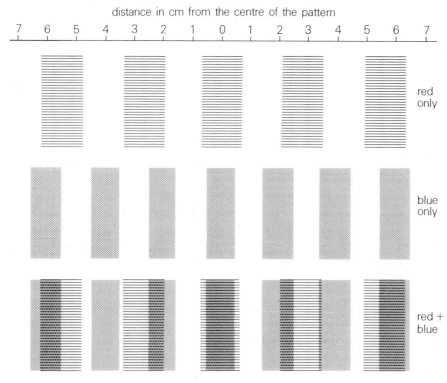

Figure 12.18 Fringes for red and blue light

Multiple slit interference

In Young's slits experiment, light was passed through two very narrow slits and, because the slits were so narrow, the amount of light energy passing through was small and the interference fringes were not very bright. A brighter pattern can be obtained by using a greater number of slits.

In Young's experiment we consider rays from the two slits and how they interfere when they meet at the screen. Since the distance d between the slits (less than 10^{-3} m) is very small compared with the distance D from the slits to the screen (usually several metres), two rays to any point on the screen will be effectively parallel, Figure 12.19.

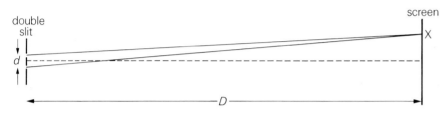

Figure 12.19 Two rays from the slits to a point on the screen

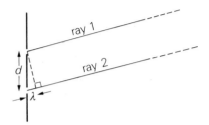

Figure 12.20 Path difference for first fringe away from the centre of the pattern

The condition for constructive interference is that the path difference between these rays is a whole number of wavelengths. For the first fringe away from the centre of the pattern, the path difference between the rays is one wavelength, Figure 12.20.

When this is the case, the waves in the two rays arrive in phase at the screen, Figure 12.21, and they produce a bright fringe by constructive interference.

Consider now what happens when a third slit is introduced at a distance d from one of the other slits. Again, because D is very large compared with d, the rays from the slits to a point on the screen will be effectively parallel. If the path difference between ray 1 and ray 2 is 1λ, the path difference between ray 2 and ray 3 will also be 1λ, Figure 12.22.

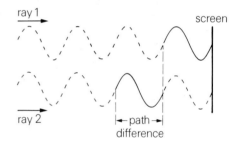

Figure 12.21 Two waves with a path difference of one wavelength

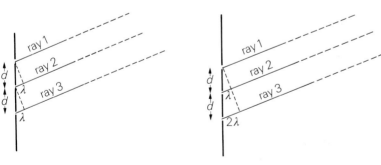

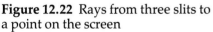

Figure 12.22 Rays from three slits to a point on the screen

Figure 12.23 Path differences between rays 2 and 3 and ray 1

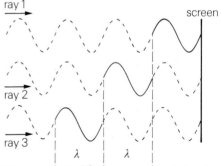

Figure 12.24 Three waves arriving in phase

This means that the path difference between ray 1 and ray 3 is 2λ, Figure 12.23, and the wave in ray 3 will also arrive in phase with ray 1, Figure 12.24.

Thus the addition of a third slit has the effect of adding further to the constructive interference producing the bright fringe. The location of the fringe is the same as it was for two slits but it will now be brighter.

The argument can be extended for any number of slits providing they are all equally spaced. If the distance between the centres of adjacent slits is d, the distance from the slits to the screen is D and the wavelength of the light is λ, then the fringe spacing Δx is given by

$$\Delta x = \frac{\lambda D}{d}$$

and the distance from the centre of the pattern is given by

$$x = \frac{n\lambda D}{d} \qquad \text{where } n = 0,1,2,3...$$

The central fringe for which $n = 0$ is called the zero order maximum. The fringe for which $n = 1$ is called the first order maximum, that for which $n = 2$ the second order maximum, and so on.

The diffraction grating is the best and most useful example of a multiple slit used for producing an interference pattern. The diffraction grating is made by a machine cutting very fine, equally spaced grooves on the surface of a glass plate. The light is diffracted by this series of grooves. Normal diffraction gratings may have between 10000 and 20000 lines per inch (about 400 to 800 lines per mm).

12.3 Diffraction grating and the spectrometer

The spectrometer is an instrument which uses either a prism or a diffraction grating to separate light into its different colours. Figure 12.25 shows a photograph of a spectrometer and Figure 12.26 shows a diagram of how it is set up using a diffraction grating.

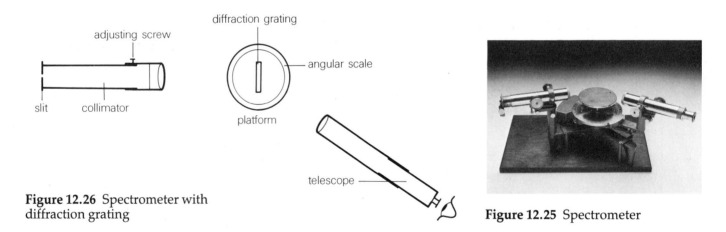

Figure 12.26 Spectrometer with diffraction grating

Figure 12.25 Spectrometer

The collimator is used to form a narrow beam of parallel light. A lamp is placed near the slit in the collimator. The length of the collimator is adjusted so that the slit falls in the focal plane of the collimator lens and this results in the light from the slit emerging as parallel rays from the collimator lens. In practice the telescope is first focused on a distant object so as to receive parallel rays. The telescope is then lined up on the collimator and the collimator length is adjusted until a clear sharp image of the slit is observed through the telescope. Since the telescope is adjusted to focus on parallel rays, the fact that the slit is sharply focused shows that the light emerging from the collimator consists of parallel rays. Once the collimator and telescope have been adjusted in this way, the diffraction grating or prism is placed on the platform. The angular scale on the platform is used for two purposes: first in setting the platform position so that the grating is at right angles to the beam of light from the collimator, and secondly to measure the angle between the telescope and this beam of light from the collimator.

With the interference pattern produced on a screen, the measurements made on the pattern were on fringe positions and spacings. With the spectrometer, the measurements made are the angles between the beams of light forming the fringes. A simple illustration of this is given in Figure 12.27.

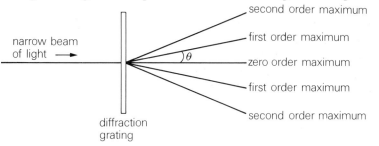

Figure 12.27 Measurement of angles between maxima

An equation can be derived to relate the path difference S_2N between rays from two slits to the distance d between the two slits and the angle θ through which the rays are being diffracted, Figure 12.28.

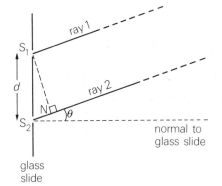

Figure 12.28 Path difference and angle of diffraction

In the triangle, $S_1\hat{N}S_2 = 90°$
$$S_1\hat{S}_2N + S_2\hat{S}_1N = 90°$$
$$S_1\hat{S}_2N + \theta = 90° \text{ (angle between the grating and the normal)}$$
$$S_2\hat{S}_1N = \theta$$

In Figure 12.28, the distance S_2N is the path difference between the two rays. The triangle S_1S_2N is shown in Figure 12.29 with the path difference S_2N, the distance d between the two slits, and the angle θ.

From Figure 12.29

$$\sin\theta = \frac{\text{path difference}}{d}$$

$\Rightarrow$ path difference $= d\sin\theta$

The condition for constructive interference is:

path difference $= n\lambda$ where $n = 0,1,2,3,...$

$\Rightarrow$ $d\sin\theta = n\lambda$

$\Rightarrow$ $\sin\theta = \dfrac{n\lambda}{d}$

For constructive interference:

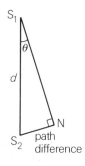

Figure 12.29 Path difference between two rays

$$\sin\theta = \frac{n\lambda}{d}$$

$\theta =$ angle of diffraction

$n =$ order of the maximum

$\lambda =$ wavelength of the light

$d =$ distance between the centres of the slits.

While this equation has been derived for two slits, it has already been shown that the same conditions apply for any number of slits providing they are equally spaced.

A spectrometer is usually used to measure the wavelength of light; d is known for the grating and the angle θ is measured: θ is the angle through which the telescope is rotated in going from the central maximum ($n = 0$) to the first order maximum ($n = 1$). This angle is greater for light of longer wavelengths, Figure 12.30.

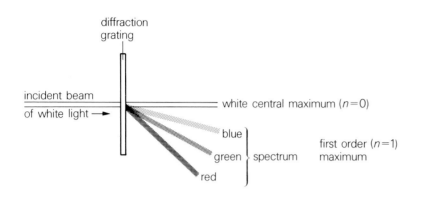

Figure 12.30 First order maxima for different colours

For the central maximum, $n = 0$

$\Rightarrow \quad \sin \theta = 0$

$\Rightarrow \qquad \theta = 0$

This is independent of wavelength, and is therefore the same for light of all wavelengths. Thus, if the incident beam is white light, the central maximum will be white.

For the first order maximum

$$\sin \theta = \frac{n\lambda}{d}$$

The wavelength, λ_b of blue light is less than the wavelength λ_r of red light. For red light the first order maximum will be at an angle θ_r such that

$$\sin \theta_r = \frac{\lambda_r}{d}$$

For blue light the first order maximum will be at an angle θ_b such that

$$\sin \theta_b = \frac{\lambda_b}{d}$$

$$\lambda_b < \lambda_r$$

$\Rightarrow \sin \theta_b < \sin \theta_r$

$\Rightarrow \qquad \theta_b < \theta_r$

Example 2

A parallel beam of white light consisting of light of wavelengths from 4×10^{-7} m (violet) to 7×10^{-7} m (red), is passed through a diffraction grating of 10 000 lines per inch (1 inch = 2.54 cm).

Describe the first order maximum and calculate the angle between the extremes of that maximum.

The first order maximum will be a spectrum of colours ranging from violet nearer the central fringe to red at the outer limit.

diffraction grating has 10 000 lines per 2.54 cm

distance between the lines $= \dfrac{2.54}{10\,000}$ cm

$$d = 2.54 \times 10^{-4}\,\text{cm}$$
$$d = 2.54 \times 10^{-6}\,\text{m}$$

For the first order maximum $\quad \sin\theta = \dfrac{\lambda}{d}$

For violet $\qquad\qquad\qquad \lambda_v = 4 \times 10^{-7}$

$\Rightarrow \qquad\qquad\qquad \sin\theta_v = \dfrac{4 \times 10^{-7}}{2.54 \times 10^{-6}}$

$\Rightarrow \qquad\qquad\qquad \sin\theta_v = 0.157$

$\Rightarrow \qquad\qquad\qquad \theta_v = 9.1°$

For red $\qquad\qquad\qquad \lambda_r = 7 \times 10^{-7}$

$\Rightarrow \qquad\qquad\qquad \sin\theta_r = \dfrac{7 \times 10^{-7}}{2.54 \times 10^{-6}}$

$\Rightarrow \qquad\qquad\qquad \sin\theta_r = 0.276$

$\Rightarrow \qquad\qquad\qquad \theta_r = 16.0°$

$\Rightarrow \qquad\qquad\qquad \theta_r - \theta_v = 16.0 - 9.1$

$\Rightarrow \qquad\qquad\qquad \theta_r - \theta_v = 6.9°$

The angle between the extremes of the first order maximum is 6.9°

Units for wavelength

Like that for any length or distance, the SI unit for wavelength is the metre. However the wavelength of light is very small, ranging from about 4×10^{-7} m to 7×10^{-7} m. Other smaller units are often used to express wavelengths of light and other electromagnetic radiations with small wavelengths. Three commonly used units are the micron (μ), the ångström (Å) and the nanometre (nm).

$$1\,\mu = 10^{-6}\,\text{m}$$
$$1\,\text{Å} = 10^{-10}\,\text{m}$$
$$1\,\text{nm} = 10^{-9}\,\text{m}$$

Example 3

A lamp gives off an intense monochromatic green light. When passed through a diffraction grating with 18 000 lines per inch, this light gives a first order maximum at an angle of 23°.
a) What is the wavelength of the light?
b) Express this wavelength in nanometres, ångströms and microns.

a) For the first order maximum, $\quad \sin\theta = \dfrac{\lambda}{d}$

The grating has 18 000 lines per 2.54 cm

$$d = \dfrac{2.54 \times 10^{-2}}{18\,000} = 1.4 \times 10^{-6}$$

$\Rightarrow \sin 23° = \dfrac{\lambda}{1.4 \times 10^{-6}}$

$\Rightarrow \qquad \lambda = 1.4 \times 10^{-6} \sin 23° = 1.4 \times 10^{-6} \times 0.39 = 5.46 \times 10^{-7}$

Wavelength of the light $= 5.46 \times 10^{-7}$ m

b) $\lambda = 5.46 \times 10^{-7}\,\text{m}$

$\Rightarrow$ $\lambda = 546 \times 10^{-9}\,\text{m}$

$\Rightarrow$ $\lambda = 546\,\text{nm}$

$\Rightarrow$ $\lambda = 5460 \times 10^{-10}\,\text{m}$ $\Rightarrow$ $\lambda = 5460\,\text{Å}$

$\Rightarrow$ $\lambda = 0.546 \times 10^{-6}\,\text{m}$ $\Rightarrow$ $\lambda = 0.546\,\mu$

12.4   Formation of spectra

A diffraction grating can be used to split light into separate colours and this is the result of an interference pattern being formed. As was seen in chapter 9, a prism may also be used to split light into separate colours. The prism refracts different colours by different amounts. A prism can be used in place of the diffraction grating on a spectrometer.

With a diffraction grating, it is light of longer wavelength that is seen to make a greater angle with the path of the incident ray. This is because the fringe spacing of the interference pattern is greater for longer wavelengths. With a prism, it is the light with the shorter wavelength that makes the greater angle because light of shorter wavelength is refracted more.

12.5   X-ray diffraction

Light produces an interference pattern when it passes through a diffraction grating. The spacing between the lines of the grating is of the same order of magnitude as the wavelength of light. The spacing between atoms in a solid is of the same order of magnitude as the wavelength of X-rays. X-rays, like light, are part of the electromagnetic spectrum, but they have a shorter wavelength than light. When X-rays are passed through a crystalline solid, a diffraction pattern is formed, Figure 12.31, and this pattern can be analysed to give information about the arrangement and spacing of the atoms in the solid.

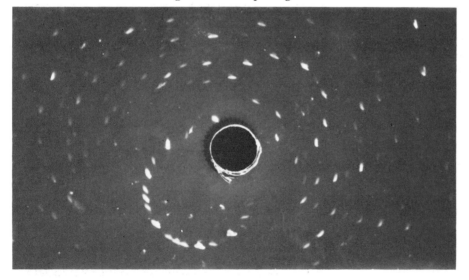

Figure 12.31 X-ray diffraction pattern

Summary

When light is passed through two very narrow slits very close together (Young's slits), it forms an interference pattern.

The fringe separation Δx in Young's Slits Experiment is given by:

$$\Delta x = \frac{\lambda D}{d}$$

where λ = wavelength of the light
D = distance from slits to screen
d = distance between the slits.

A diffraction grating consists of a large number of fine, equally spaced, parallel grooves on the surface of a glass plate, and produces an interference pattern similar to that of two slits, but of greater brightness.

A spectrometer consists of:

a collimator to produce a narrow beam of parallel light;

a platform on which to mount a prism or diffraction grating;

a telescope to locate and view the interference fringes.

For a diffraction grating, the angle θ through which the rays are diffracted for a bright fringe is given by;

$$\sin \theta = \frac{n\lambda}{d}$$

where n = order of the maximum
λ = wavelength of the light
d = distance between the centre of the slits.

Common units for the wavelength of light are:

1 micron (μ) = 10^{-6} m

1 angstrom (Å) = 10^{-10} m

1 nanometre (nm) = 10^{-9} m

X-ray diffraction patterns can be produced by passing X-rays through a crystalline solid, and can be used to give information about the arrangement and spacing of atoms in the solid.

Problems

1   Which are diffracted least: waves of shorter wavelength or waves of longer wavelength? How does this account for the fact that it was some time before people would accept Huygens' wave theory of light?

2   At the start of the eighteenth century there were two main theories of the nature of light: the Wave Theory and the Corpuscular Theory. What was the experimental evidence, demonstrated by Thomas Young, which strongly supported the Wave Theory?
Explain, with the aid of a diagram, how you would set up apparatus to demonstrate this experimental evidence in the laboratory.

3   In an experiment to demonstrate the interference of light, red light was passed through two narrow slits ruled in black paint on a glass slide. A pattern of red interference fringes was produced on a screen.
State what effect each of the following changes would have on the spacing between the interference fringes:
a) Replacing the red light source by a blue light source.
b) Replacing the glass slide by another on which the two narrow slits are closer together.
c) Increasing the distance between the screen and the glass slide.

4   Light from a sodium lamp is passed through two narrow slits to produce an interference pattern on a screen. When the distance from the slits to the screen is 2 m and the spacing between the two slits is 0·2 mm, the interference fringes have a spacing of 6 mm between their centres. What is the wavelength of the sodium light?

5   Light of wavelength 4.5×10^{-7} m is passed through two narrow slits, 0.3 mm apart and an interference pattern is produced on a screen 3 m from the slits. What is the distance from the centre of the interference pattern to the eighth fringe of the interference pattern?

6   White light is passed through a filter which absorbs green light. After passing through the filter the light is passed through two narrow slits so that an interference pattern is produced on a screen. The interference pattern is a mixture of red and blue fringes. The central fringe of the pattern is a mixture of red and blue light. The next case in which the centres of the red and blue fringes coincide is where the third red fringe from the centre coincides with the fourth blue fringe from the centre. If the wavelength of the red light is 6.4×10^{-7} m, what is the wavelength of the blue light?

7   A spectrometer has a collimator, a telescope and a platform on which a prism or a diffraction grating is mounted.
State briefly what is the function of:
a) the collimator;
b) the telescope;
c) the prism or diffraction grating.

8   White light can be separated to form a continuous spectrum by passing it through a spectrometer using either a prism or a diffraction grating on the platform. What is the main difference between the angles at which the different colours are observed when using a prism or a diffraction grating?

9   Using a spectrometer, light of wavelength 6.25×10^{-7} m is passed through a diffraction grating with 600 lines per mm. What is the angle between the first order maximum and the second order maximum?

10   Using a spectrometer, white light of wavelength from 4.5×10^{-7} m to 7.5×10^{-7} m is passed through a diffraction grating with 500 lines per mm.
a)   What colour is the central (zero order) maximum?
The first order maximum consists of a range of colours.
b)   What is the angle between the extremes of the first order maximum?
c)   Which colour is nearer the central order maximum?

11   Express the following lengths in metres:
a) $0.7\,\mu$;   **b)** $25.0\,\mu$;   **c)** $7000\,\text{Å}$;   **d)** $400\,\text{nm}$.

12   What is 6.6×10^{-7} m when expressed in:
a) microns;   **b)** nanometres;   **c)** Ångstroms?

13   Why are X-rays rather than light used to produce a diffraction pattern of a crystalline solid?

14   A laser is a device which produces a narrow beam of monochromatic light. One type of laser produces red light.
a)   Light from this laser is allowed to strike a blackened glass plate on which there are two narrow parallel slits. The light emerging from the slits is viewed on a screen placed a distance from the slits as shown in Figure 1.

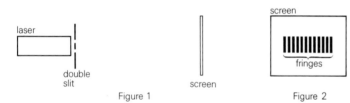

Figure 1   Figure 2

On viewing the screen a series of equally spaced fringes is observed, Figure 2.

i) Explain how the fringe pattern is produced.
ii) Suggest how the pattern observed on the screen would be affected if:
a) blue light from a second laser replaced the beam of red light.
b) the beam of red light was allowed to shine on a pair of slits tapering as shown in Figure 3, the beam being gradually moved from X to Y.

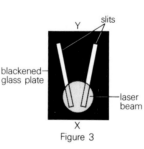

Figure 3

15   A pupil holds a double slit in front of his eye and looks at a tungsten filament lamp with a scale immediately behind it.

a)   A red filter is placed in front of the lamp. Describe what he sees and explain in terms of waves how this arises.
b)   The red filter is then replaced by a blue one. Explain any difference in fringe separation with blue and with red light.
c)   Explain why the fringes have coloured edges if no filter is used.
d)   With the red filter in place, the pupil estimates the apparent separation of the bright fringes to be 5.0 mm when $L = 2.0$ m. If the slit separation is 0.25 mm, what is the wavelength of the light passing through the filter?
You may use the relationship $\Delta x = \lambda L/d$ for the apparent separation Δx of the fringes.   *SCEEB*

16   The sketch illustrates one method of producing interference fringes.

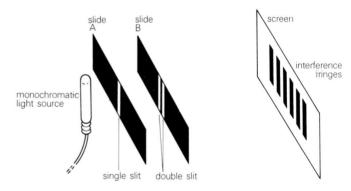

a)   In each case, state the effect on the interference fringes of
i)   moving the screen further away from slide B;
ii)   replacing slide B by another similar slide on which the two slits are closer together;
iii)   replacing the monochromatic light source by another monochromatic light source emitting light of longer wavelength.
b)   The monochromatic light source is replaced by a tungsten filament lamp and slide B is replaced by a diffraction grating.
i)   Explain in terms of waves why the red colour of the observed spectrum is seen further away from the central band than the blue colour.
ii)   Explain why the central band, known as the zero order spectrum, is white.   *SCEEB*

13 A particle model

13.1   Introduction

Since the time of the ancient Greeks, it has been suspected that all substances are built up from small particles. It is only within the last two centuries that experimental evidence which supports this view has been obtained.

Robert Brown observed pollen dust which was suspended in water and he found that the dust was seen to be moving in a random manner. The same effect can be seen with smoke in an air cell, Figure 13.1. Smoke consists of small specks of ash which are large enough to be seen; the jerky movement which is observed is due to the particles of air colliding with the specks of ash.

We can explain in a fairly simple way the difference between solids, liquids and gases by assuming that they are made up of particles. Gases are easy to compress, but solids and liquids require very large forces to compress them by even a small amount. We believe that this is because the particles in solids and liquids are packed tightly together, but in a gas they are relatively far apart. In liquids, the particles are closely packed but can change position so that there is no fixed shape. In a gas, the particles are further apart and move about colliding with each other and with the walls of the container.

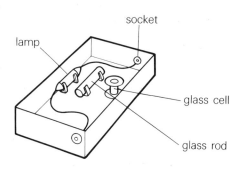

Figure 13.1 Brownian motion apparatus

13.2   Particle separation

When a solid melts, there is very little change in volume. However, when a liquid is changed to a gas, there is a large increase in volume.

Two experiments can provide us with an estimate of the change in volume when a liquid changes to a gas and when a solid changes to a gas.

Liquid to gas

The plunger of a calibrated syringe is pushed to the bottom to exclude all air and the nozzle is sealed with a rubber cap, Figure 13.2. A small measured volume of water is then injected through the rubber seal by means of a hypodermic syringe. The large calibrated syringe is then immersed in a beaker of salt solution which is boiling at a temperature just above 100°C. The

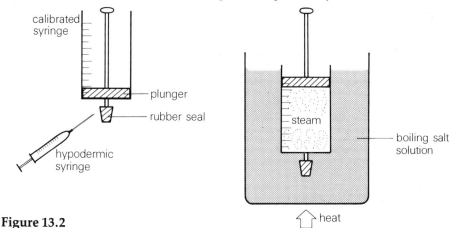

Figure 13.2

sample of water turns into steam which pushes back the plunger. The volume of steam produced is read off the scale of the syringe. It is found that the water produces a volume of steam that is about 1600 times greater than the volume of water.

Solid to gas

A small test tube containing about $1\,cm^3$ of solid carbon dioxide is connected by tubing to an inverted gas jar, Figure 13.3. The gas jar is calibrated for measuring volume and is initially filled with water. As the solid carbon dioxide vaporizes, it is collected in the gas jar. It is found that the volume of gas produced is about 800 times greater than the volume of solid.

Results from experiments such as these indicate that the volume occupied by particles of a gas is of the order of one thousand times greater than when in the liquid or solid state.

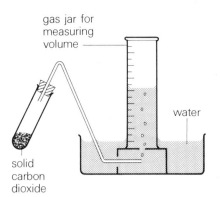

Figure 13.3

Density

Although the volume of the particles has increased, the mass of the particles, whether in the gaseous or the liquid state, has remained unchanged. If we take equal volumes of the liquid and of the gas, their mass would be very different.

The mass per unit volume of a substance is called its **density** ρ.

$$\text{density } \rho = \frac{\text{mass}}{\text{volume}}$$

$$\Rightarrow \qquad \rho = \frac{M}{V}$$

The SI units for density are $kg\,m^{-3}$.

In general, solids and liquids do expand a little when heated: the same mass will then occupy a slightly larger volume, and the density will be smaller. An accurate value of the density should also state the temperature at which it was measured.

A change in temperature or in pressure can produce a large change in the volume of gas. The value of a gas density must state the temperature and the pressure at which it was measured. The conditions chosen are usually a temperature of 0°C and a pressure of $1.01 \times 10^5\,Pa$: this is known as **Standard Temperature and Pressure** (s.t.p.).

Table 1 shows the densities of some solids, liquids and gases.

solids and liquids	density at 20°C /$kg\,m^{-3} \times 10^3$	gases	density at s.t.p. /$kg\,m^{-3}$
cork	0.25	hydrogen	0.09
olive oil	0.92	helium	0.18
water	1.00	nitrogen	1.25
naphthalene	1.15	air	1.29
perspex	1.19	oxygen	1.43
glycerol	1.26	carbon dioxide	1.98
aluminium	2.70		
glass	3.00		
iron	7.86		
silver	10.50		
lead	11.40		
mercury	13.60		
gold	19.30		

Table 1 Densities

Example 1

A volume of $5 \times 10^{-4}\,m^3$ of alcohol has a mass of $0.4\,kg$.

What is the density of the sample?

$$\rho = \frac{M}{V} \qquad\qquad M = 0.4\,kg$$
$$V = 5 \times 10^{-4}\,m^3$$

$$\Rightarrow \quad \rho = \frac{0.4}{5 \times 10^{-4}}$$

$$\Rightarrow \quad \rho = 800$$

The density of alcohol is $800\,kg\,m^{-3}$

Example 2

A sample of oxygen with a density of $1.43\,kg\,m^{-3}$ at s.t.p. occupies a volume of $5.00 \times 10^{-3}\,m^3$ at s.t.p. What is the mass of the sample?

$$\rho = \frac{M}{V} \qquad\qquad \rho = 1.43\,kg\,m^{-3}$$
$$V = 5.00 \times 10^{-3}\,m^3$$

$$\Rightarrow 1.43 = \frac{M}{5.00 \times 10^{-3}}$$

$$\Rightarrow \quad M = 1.43 \times 5.00 \times 10^{-3}$$

$$\Rightarrow \quad M = 7.15 \times 10^{-3}$$

The mass of the sample is $7.15 \times 10^{-3}\,kg$

13.3   Spacing of particles

Figure 13.4 shows a model of the structure of a liquid in which it is assumed that the particles are spherical. This shows that when the particles are closely packed together, the spacing d is equal to the particle diameter d_0. If the liquid is compressed, this is resisted by a repulsion force between the particles. If the particles are pulled apart, this is resisted by an attraction force between the particles.

As we saw earlier, the change in volume which takes place when a sample of liquid changes into a gas is by a factor of the order of 1000 times. To estimate the change in spacing which will take place, it is assumed that each particle is enclosed in a cubical box, Figure 13.5. The sides of the box have a length equal to the particle diameter d_0. The volume is therefore d_0^3

When the liquid becomes a gas, the size of the particles does not change but they become more widely spaced and the volume occupied by one particle will be 1000 times greater than the original volume, Figure 13.6.

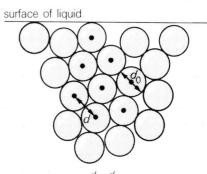

surface of liquid

Figure 13.4 $d = d_0$

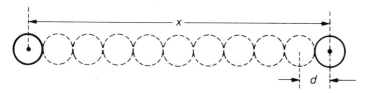

Figure 13.6

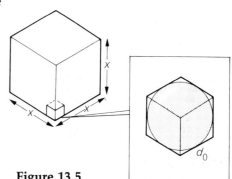

Figure 13.5

The new volume is $1000\,d_o^3$ which can be represented by a cube of side x. In this case the new volume will be x^3.

$$x^3 = 1000\,d_o^3$$
$$\Rightarrow\; x = \sqrt[3]{1000\,d_o^3}$$
$$\Rightarrow\; x = 10\,d_o$$
$$\Rightarrow\; x = 10\,d \qquad \text{since } d = d_o$$

Thus in a gas the particle spacing is about 10 times greater than that of the liquid.

13.4   Experimental determination of particle size

It is impossible to measure the size of a particle directly, but the size of an oil molecule can be estimated fairly easily. By allowing a small drop of oil to fall on to the surface of water, a circular oil film is formed on the surface. If we assume that the film spreads until minimum thickness is obtained, it is reasonable to suppose that the thickness will be equal to the size of an oil molecule. A single layer of molecules will be produced starting from a cluster of molecules in the same way that a handful of marbles thrown on to a tray will spread out into a layer one marble thick. In the experiment, a small drop of oil is suspended from a thin wire and the size of the drop is compared against a scale using a magnifying glass, Figure 13.7.

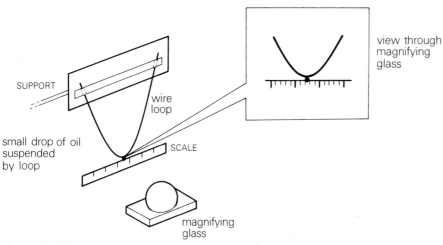

Figure 13.7

The surface of the water to be used is coated with a light powder and the drop is allowed to fall on the surface. A film of oil spreads out, pushing back the powder to form a circular disc shape. The oil continues spreading outwards until the film has achieved its minimum thickness. It is assumed that this occurs when the film is one molecule thick. This is called a monolayer. The diameter of the oil film is measured using a metre stick.

If the thickness of the film is t and its radius is R, the volume of the film (a disc) will be

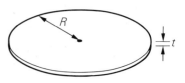

volume = area of cross section × thickness

$$V_{\text{film}} = \pi R^2 t$$

If the radius of the original drop (a sphere) was r, the volume of the drop will be

$$V_{drop} = \tfrac{4}{3}\pi r^3$$

and since the volume of the film must be the same as the drop

$$\pi R^2 t = \tfrac{4}{3}\pi r^3$$

$$\Rightarrow \quad t = \frac{4r^3}{3R^2}$$

Example 3

The molecular diameter of an oil molecule is found by dropping a small drop on to the surface of water. The radius of the oil drop is found to be 0.35×10^{-3} m and the radius of the film is 0.25 m.

Find the thickness of the film.

radius of the film $R = 0.25$ m

radius of the drop $r = 0.35 \times 10^{-3}$ m

thickness of film $t = \dfrac{4r^3}{3R^2}$

$$\Rightarrow \qquad t = \frac{4 \times (0.35 \times 10^{-3})^3}{3 \times 0.25^2}$$

$$\Rightarrow \qquad t = 9.2 \times 10^{-10}$$

Assuming the film to be a monolayer, the diameter of the molecule is 9.2×10^{-10} m.

Molecular diameters are normally given in nanometres where 1 nanometre (nm) $= 1 \times 10^{-9}$ m

The diameter of the oil molecule is therefore 0.92 nm

13.5   Mass of an atom

Our model has assumed that substances consist of particles. These particles are made up of one or more smaller units called atoms. The mass of an atom is very small: the smallest atom, hydrogen, has a mass of 1.67×10^{-27} kg. For convenience, it is usual to compare the mass of atoms with the mass of a standard atom. Originally, the hydrogen atom was taken as the standard because it was the smallest but in 1960 it was decided internationally to adopt the carbon atom $^{12}_{6}C$ as the standard. The mass of this atom was taken as 12 unified atomic mass units, normally abbreviated to 12 u. Values of atomic mass based on this standard are given in Table 2.

From Table 2 we can see that

$$\frac{\text{mass of 1 helium atom}}{\text{mass of 1 carbon atom}} = \frac{4}{12}$$

$$\frac{\text{mass of } x \text{ helium atoms}}{\text{mass of } x \text{ carbon atoms}} = \frac{4x}{12x}$$

$$\frac{\text{mass of } x \text{ helium atoms}}{\text{mass of } x \text{ carbon atoms}} = \frac{4}{12}$$

This shows that the ratio of the masses of equal numbers of atoms of different substances is equal to the ratio of their atomic masses. It follows from this that

element	atomic mass u
hydrogen	1.01
helium	4.00
lithium	6.94
carbon	12.00
nitrogen	14.01
oxygen	16.00
sodium	23.00
aluminium	27.00
silicon	28.09
calcium	40.08
chromium	52.00
iron	55.85
copper	63.54
strontium	87.62
silver	107.90
iodine	127.00
gold	197.00
radon	222.00
uranium	238.00
plutonium	244.00

Table 2

the samples contain equal numbers of atoms. From the table it can be seen that 12 g of carbon and 4 g of helium contain the same number of atoms. Similarly, 12 g of carbon and 52 g of chromium contain the same number of atoms. In general, for any substance of atomic mass x, x grams will contain the same number of atoms as 12 g of carbon. The number of atoms in 12 g of carbon is 6×10^{23} and this number is known as Avogadro's Number (N_A). The mass of any substance which contains this number of particles is called a **gram-mole**: the particles may be atoms or molecules.

Example 4

If the atomic mass of carbon is 12 u and Avogadro's Number is 6×10^{23}, find the mass of a carbon atom.

one gram-mole of carbon = 12 g

this contains 6×10^{23} atoms

$$\text{mass of 1 atom} = \frac{12}{6 \times 10^{23}}$$

$\Rightarrow \quad \text{mass of 1 atom} = 2 \times 10^{-23}$

The mass of 1 atom of carbon is 2×10^{-23} g which is 2×10^{-26} kg

Example 5

The density of helium is $0.18 \, \text{kg m}^{-3}$ at s.t.p. Find the volume of 1 mole of helium at s.t.p.

   The atomic mass of helium is 4.00 u so that 1 mole of helium has a mass of 4 grams.

$$\rho = \frac{M}{V} \qquad\qquad M = 0.004 \, \text{kg}$$
$$\rho = 0.18 \, \text{kg m}^{-3}$$

$\Rightarrow \quad 0.18 = \dfrac{0.004}{V}$

$\Rightarrow \quad V = \dfrac{0.004}{0.18}$

$\Rightarrow \quad V = 22.4 \times 10^{-3}$

The volume of 1 mole of helium at s.t.p. is $22.4 \times 10^{-3} \, \text{m}^3$

13.6   Behaviour of gases

Pressure is defined in the following way

$$\text{pressure} = \frac{\text{force acting at right angles to an area}}{\text{area over which the force acts}}$$

The SI unit of pressure is the pascal where $1 \, \text{Pa} = 1 \, \text{N m}^{-2}$

When the pressure on a solid or liquid is altered there is hardly any change in its volume. With a gas a very large change of volume is possible.

Robert Boyle investigated the relationship between the pressure p and the volume V for a fixed mass of gas at constant temperature. The apparatus shown in Figure 13.9 can be used for this investigation. The results when plotted give Figure 13.10. If $1/V$ is plotted against p, the straight line graph in Figure 13.11 is obtained.

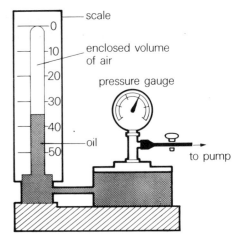

Figure 13.9 Boyle's Law apparatus

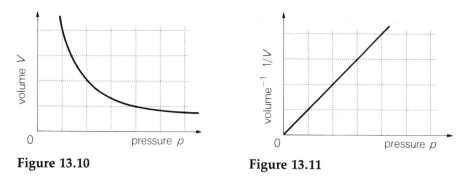

Figure 13.10   **Figure 13.11**

This illustrates that

$$p \propto \frac{1}{V} \text{ (mass and temperature constant)}$$

$$pV = \text{constant (mass and temperature constant)}$$

This is known as **Boyle's Law** which can be written

$$p_1 V_1 = p_2 V_2 \quad \text{where } V_1 \text{ is the volume at pressure } p_1$$
$$\text{and} \quad V_2 \text{ is the volume at pressure } p_2$$

Example 6

A cylinder contains $0.2\,\text{m}^3$ of oxygen at a pressure of $3 \times 10^5\,\text{Pa}$. What volume will this oxygen occupy at a pressure of $1 \times 10^5\,\text{Pa}$ if the temperature is unchanged?

$$p_1 V_1 = p_2 V_2 \qquad V_1 = 0.2\,\text{m}^3; \qquad V_2 = \text{final volume}$$
$$p_1 = 3 \times 10^5\,\text{Pa}; \qquad p_2 = 1 \times 10^5\,\text{Pa}$$

$$\Rightarrow 3 \times 10^5 \times 0.2 = 1 \times 10^5 \times V_2$$

$$\Rightarrow \qquad V_2 = \frac{3 \times 10^5 \times 0.2}{1 \times 10^5}$$

$$\Rightarrow \qquad V_2 = 0.6$$

The new volume will be $0.6\,\text{m}^3$

The pressure exerted by a gas is caused by the bombardment of the sides of the container by the gas molecules. If the temperature is increased, the kinetic energy of the molecules will also increase and the pressure will be greater because of more frequent collisions by particles colliding with greater energy.

The relationship between pressure and temperature can be investigated by heating a fixed mass of gas at constant volume over a range of temperatures, Figure 13.12, and measuring the corresponding pressures.

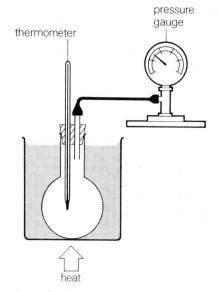

Figure 13.12

Figure 13.13   **Figure 13.14**

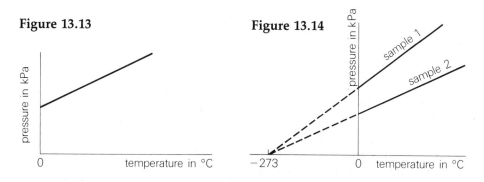

A graph of the results from such an experiment is shown in Figure 13.13. If the experiment is repeated with a larger flask, a different set of results is obtained. When the results of both experiments are plotted and the lines extended backwards, they both intersect the temperature axis at $-273°C$, Figure 13.14.

In general this is true for any graph of pressure against temperature for a fixed mass of gas. This temperature, $-273°C$, is known as **absolute zero** and is the starting point for the absolute temperature scale. When the pressure is plotted against the absolute temperature in kelvin (K), the pressure line passes through the origin as shown in Figure 13.15.

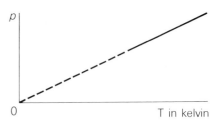

Figure 13.15

This means that

$p \propto T$ (mass and volume constant)

$\Rightarrow \quad \dfrac{p}{T} = $ constant

where T is the absolute temperature measured in kelvin.

This is known as the **Pressure Law**

$$\dfrac{p}{T} = \text{constant}$$

$\Rightarrow \quad \dfrac{p_1}{T_1} = \dfrac{p_2}{T_2}$ where p_1 is the pressure at absolute temperature T_1 and p_2 is the pressure at absolute temperature T_2

If the pressure of a gas is kept constant and the temperature raised, the volume increases. The relationship can be investigated using the apparatus shown in Figure 13.16 A straight line graph is obtained, Figure 13.17. This illustrates **Charles's Law** which states that the volume is proportional to the absolute temperature if the pressure and mass is constant.

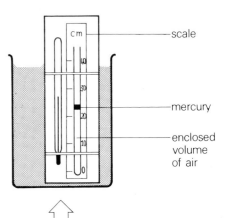

heat

Figure 13.16

$V \propto T$ (mass and pressure constant)

$\Rightarrow \quad \dfrac{V}{T} = $ constant

where T is the absolute temperature measured in kelvin.

$\Rightarrow \quad \dfrac{V_1}{T_1} = \dfrac{V_2}{T_2}$ where V_1 is the volume at temperature T_1 and V_2 is the volume at temperature T_2

Combining these laws gives the **Combined Gas Equation** which (for a constant mass of gas) relates the absolute temperature T to the pressure and the volume.

$$pV = \text{constant} \times T$$

The expression for the constant is given by:

$$\text{constant} = \dfrac{pV}{T}$$

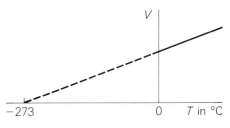

Figure 13.17

The value of this constant will depend on the mass of gas under consideration.

Example 7

A sample of oxygen occupies $5\,m^3$ at a pressure of $1 \times 10^5\,Pa$ and a temperature of $27°C$. What volume does it occupy at a pressure of $4 \times 10^5\,Pa$ and a temperature of $-3°C$?

$$\dfrac{p_1 V_1}{T_1} = \dfrac{p_2 V_2}{T_2}$$

$\Rightarrow \quad \dfrac{1 \times 10^5 \times 5}{300} = \dfrac{4 \times 10^5 \times V_2}{270}$

$\Rightarrow \quad V_2 = 5 \times \dfrac{1 \times 10^5}{4 \times 10^5} \times \dfrac{270}{300} = 1.13$

$V_1 = 5.00\,m^3 \quad V_2 = ?$
$p_1 = 1.00 \times 10^5\,Pa$
$p_2 = 4.00 \times 10^5\,Pa$
$T_1 = 273 + 27 = 300\,K$
$T_2 = 273 - 3 = 270\,K$

The volume of the sample is $1.13\,m^3$

13.7 Kinetic Theory of gases

The gas laws were derived from experimental evidence of the behaviour of gases. We shall develop a kinetic model of gases which can assist us to describe this behaviour of gases.

The model is based on the assumption that all gases consist of distinct particles which are completely separate and moving in many directions with various speeds. We call this random motion. The particles collide with each other and with the walls of the container.

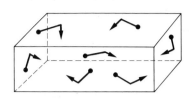

Figure 13.18

The following assumptions are also made.

1 there is a large number of particles in the container

2 the particles move in random directions

3 the particles have a range of speeds

4 the actual volume of the particles is negligible compared to the volume of the container

5 all collisions are perfectly elastic

6 the particles do not exert forces on each other except during collisions

7 the time of contact spent by the particles during collisions is negligible

8 the particles obey Newton's Laws

We shall assume that the gas is enclosed in a container, length l, height h and breadth b, the edges of which lie along the axes O_x, O_y and O_z as indicated in Figure 13.19. There is a large number of particles in the container and these particles collide with the walls at different angles. The velocity of one particle P can be resolved into three components v_x, v_y and v_z parallel to the sides of the container, Figure 13.20.

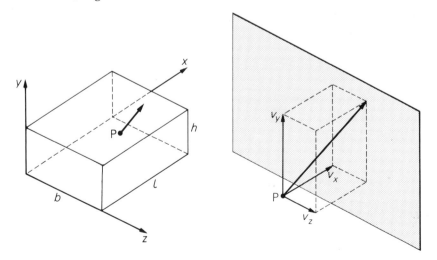

Figure 13.19   **Figure 13.20**

We shall first consider the x-component v_x of the velocity so that we imagine the particle bouncing backwards and forwards between two opposite faces, all collisions taking place at right angles, Figure 13.21. The mass of the particle is m.

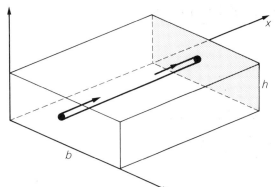

Figure 13.21

Figure 13.22

Since the collisions are perfectly elastic, the particle will rebound with the same speed in the opposite direction. If we take left-to-right as the positive direction for vectors,

initial momentum of particle $= + mv_x$

final momentum of particle $= - mv_x$

the change of momentum will be given by

change of momentum = final momentum − initial momentum

$$= - mv_x - (+ mv_x)$$

$$= - 2mv_x$$

If we suppose that the molecule takes time t to travel the distance $2l$,

$$\text{speed} = \frac{\text{distance}}{\text{time}} \quad \Rightarrow \quad v_x = \frac{2l}{t}$$

$$t = \frac{2l}{v_x}$$

As shown in Figure 13.23 we can see that, regardless of the starting position of the particle, there will be one collision per wall during this time interval t.

The magnitude of the force acting on the particle between collisions is given by

$$\text{force on particle} = \frac{\text{change of momentum of particle per collision}}{\text{time between collisions}}$$

$$= \frac{-2mv_x}{\frac{2l}{v_x}} = \frac{-mv_x{}^2}{l}$$

This is the force exerted on the particle by the wall, so that by Newton's Third Law, the force exerted by the particle on the wall will be:

$$\text{average force on wall} = \frac{+ mv_x^2}{l}$$

Figure 13.23

There are N particles in the container and each will have its own particular x-component of velocity.

The average value $\overline{v_x^2}$ for the square of the velocity of N particles is given by

$$\overline{v_x^2} = \frac{v_{x_1}^2 + v_{x_2}^2 + v_{x_3}^2 + \dots + v_{x_N}^2}{N}$$

In Figure 13.21, the force F acting on the shaded wall due to all the particles present in the container is given by

$$F = \text{number of particles} \times \text{average force}$$

$$= N \times \frac{m\overline{v_x^2}}{l}$$

If the area of the shaded wall is $h \times b$, the pressure p_x on this wall is given by

$$p_x = \frac{\text{force acting at right angles to an area}}{\text{area of surface}} = \frac{Nm\overline{v_x^2}}{l} \div hb$$

$$= \frac{Nm\overline{v_x^2}}{hbl}$$

$$= \frac{Nm\overline{v_x^2}}{V} \qquad \text{where } V = \text{volume of the container}$$

The expression for pressure has been developed using the x-component of velocity.

The actual velocity of the molecule is c as shown in Figure 13.24. In this vector diagram the lengths represent velocities so that

$\overrightarrow{OA}$ = length of vector v_x
$\overrightarrow{AD}$ = length of vector v_z
$\overrightarrow{DE}$ = length of vector v_y

Applying Pythagoras' Theorem to triangle OAD

$$\overrightarrow{OD}^2 = \overrightarrow{OA}^2 + \overrightarrow{AD}^2 \qquad \Rightarrow \qquad \overrightarrow{OD}^2 = v_x^2 + v_z^2$$

and similarly in triangle ODC

$$\overrightarrow{OE}^2 = \overrightarrow{OD}^2 + \overrightarrow{DE}^2 \qquad \Rightarrow \qquad c^2 = v_x^2 + v_z^2 + v_y^2$$

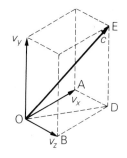

Figure 13.24

If we consider the averages of all the molecular velocities

$$\overline{c^2} = \overline{v_x^2} + \overline{v_z^2} + \overline{v_y^2}$$

For a large number of particles moving at random, there is no reason why the average component in one direction should be different from the others.

$$\overline{v_x^2} = \overline{v_y^2} = \overline{v_z^2} \qquad \text{so that } \overline{c^2} = 3\overline{v_x^2}$$

$$\Rightarrow \qquad \overline{v_x^2} = \tfrac{1}{3}\overline{c^2}$$

and the expression for pressure becomes

$$p = \frac{Nm}{V}(\tfrac{1}{3}\overline{c^2}) \qquad \Rightarrow \qquad p = \tfrac{1}{3}\frac{Nm}{V}\overline{c^2} \qquad \Rightarrow \qquad pV = \tfrac{1}{3}Nm\overline{c^2}$$

The total mass of the gas $= Nm$ where N = total number of particles

m = mass of one particle

Since density $= \dfrac{\text{mass}}{\text{volume}}$ $\rho = \dfrac{Nm}{V}$

pressure $p = \tfrac{1}{3}\rho\overline{c^2} \qquad \Rightarrow \qquad \overline{c^2} = \dfrac{3p}{\rho}$

If the pressure and density of the gas are known, we can calculate the value of $\overline{c^2}$. This term is the average of the squares of the speeds of the particles. It is called the mean square speed.

To obtain an estimation for the average speed of the particles we take the square root of the mean square speed.

$$\sqrt{\overline{c^2}} = \sqrt{\frac{3p}{\rho}}$$

The term $\sqrt{\overline{c^2}}$ is called the **root mean square speed** of the particle which is written as c_{rms}

$$c_{rms} = \sqrt{\frac{3p}{\rho}}$$

This is not the same as the average speed.

Example 8

Some gas molecules have the following individual speeds in $m\,s^{-1}$: 56, 120, 194, 250, 286, 295, 299, 301, 306, 312, 317, 405.

Find **a)** the average speed **b)** the r.m.s. speed

	speed	speed squared
	56	3136
	120	14400
	194	37636
	250	62500
	286	81796
	295	87025
	299	89401
	301	90601
	306	93636
	312	97344
	317	100489
	405	164025
total	3141	921989

$$\text{average value of speed} = \frac{3141}{12} = 262$$

$$\text{average value of speed squared} = \frac{921989}{12} \quad \Rightarrow \quad \overline{c^2} = 76832$$

$$\Rightarrow \quad c_{rms} = 277$$

The average speed is 262 m s^{-1} and the r.m.s. speed is 277 m s^{-1}

Example 9

Find the r.m.s. speed of oxygen molecules at s.t.p. if the density of oxygen is $1.43\,kg\,m^{-3}$ at s.t.p.

At s.t.p. the pressure $= 1.01 \times 10^5\,Pa$ and $p = \frac{1}{3}\rho\overline{c^2}$

$\Rightarrow$ $1.01 \times 10^5 = \frac{1}{3} \times 1.43 \times \overline{c^2}$

$\Rightarrow$ $\overline{c^2} = \dfrac{3 \times 1.01 \times 10^5}{1.43} = 2.12 \times 10^5$ $\Rightarrow$ $c_{rms} = \sqrt{2.12 \times 10^5} = 460.3$

The r.m.s. speed at s.t.p. is 460 m s^{-1}

13.8   Kinetic Theory and the combined gas equation

The equation $pV = \frac{1}{3}Nm\overline{c^2}$ can be modified to

$$pV = \frac{2}{3}N(\frac{1}{2}m\overline{c^2}) \text{ where } \frac{1}{2}m\overline{c^2} \text{ is the average kinetic energy}$$
$$\text{of a molecule.}$$

For a fixed mass of gas, the number of molecules N is constant.

$$\Rightarrow \qquad pV \propto \text{average kinetic energy}$$

If we assume that the average kinetic energy is directly proportional to the absolute temperature T

$$pV \propto T$$
$$\Rightarrow \qquad pV = \text{constant} \times T$$

This relationship is derived from considerations of kinetic theory and agrees with the experimentally obtained result.

Example 10

If the r.m.s. speed of neon molecules is $581 \, \text{m s}^{-1}$ at s.t.p., find the r.m.s. speed at 100°C and the same pressure.

Let $\sqrt{\overline{c_1^2}}$ be the r.m.s. speed at temperature T_1

and $\sqrt{\overline{c_2^2}}$ be the r.m.s. speed at temperature T_2

Since the average kinetic energy is directly proportional to the absolute temperature,

$$\frac{1}{2}m\overline{c_1^2} \propto T_1 \quad \text{and} \quad \frac{1}{2}m\overline{c_2^2} \propto T_2$$

$$\Rightarrow \frac{\frac{1}{2}m\overline{c_1^2}}{\frac{1}{2}m\overline{c_2^2}} = \frac{T_1}{T_2}$$

$$\Rightarrow \frac{581^2}{\overline{c_2^2}} = \frac{273}{373}$$

$$\Rightarrow \overline{c_2^2} = \frac{373}{273} \times 581^2$$

$$\Rightarrow \sqrt{\overline{c_2^2}} = \sqrt{4.61 \times 10^5} = 679$$

The r.m.s. speed at 100°C is 679 m s^{-1}

Summary

$$\text{pressure} = \frac{\text{force acting normal to an area}}{\text{area over which force acts}}$$

$$\text{density} = \frac{\text{mass}}{\text{volume}}$$

The combined gas equation is $\dfrac{pV}{T} = \text{constant}$

The spacing of molecules in solids and liquids is approximately one tenth the spacing in a gas

The kinetic theory is used to derive the equations:

$$pV = \tfrac{1}{3}Nm\overline{c^2}$$
$$p = \tfrac{1}{3}\rho\overline{c^2}$$

The average kinetic energy of the molecules of a gas is directly proportional to the absolute temperature.

Problems

1   A syringe has a piston with cross-sectional area $2\,\text{cm}^2$. The piston is pushed with a force of 12N. Calculate the pressure.

2   A drawing pin has a sharp point with area $0.01\,\text{mm}^2$. What is the pressure exerted by the point when the head is pushed with a force of 8N?

3   Using a kinetic model of matter, distinguish between solids, liquids and gases.

4   Explain how you would estimate the density of air at atmospheric pressure and room temperature.

5   Kinetic theory predicts the equation $pV = \tfrac{1}{3}Nm\overline{c^2}$
Write down the meaning of each of the symbols in this equation.

6   The diameter of an oil drop is found to be $6.5 \times 10^{-4}\,\text{m}$. Calculate the volume of this drop. This drop is allowed to fall on to the surface of water. It spreads out forming a circular film of diameter $0.43\,\text{m}$. From these figures, estimate the length of an oil molecule.

7   At s.t.p. $22.4 \times 10^{-3}\,\text{m}^3$ of any gas contain 6×10^{23} molecules. What is the average volume occupied by one molecule? If $0.002\,\text{kg}$ of hydrogen contain 6×10^{23} molecules, calculate a value for the density of hydrogen at s.t.p.

8   'When the absolute temperature of an ideal gas is doubled, the average molecular speed is doubled and the pressure is also doubled. Criticise this statement.

9   In an experiment the temperature of a fixed mass of gas is kept constant. The pressure is altered and various readings of pressure and volume are taken. These are listed in the table

pressure (k Pa)	101	116	122	135	180	210	250
volume (cm³)	45	39	37	34	25	22	18

Plot a graph of pressure against volume.
What is the relationship between pressure and volume?

10   $1 \times 10^{-3}\,\text{m}^3$ of gas has a pressure of 4 kPa at a temperature of 20°C. If the pressure increases to 9 kPa and the temperature decreases to -15°C determine the new volume.

11   A fixed mass of gas is kept at constant temperature but the pressure is increased from $1.01 \times 10^5\,\text{Pa}$ to $3 \times 10^5\,\text{Pa}$. If the original volume was $0.2\,\text{m}^3$, determine the final volume.

12   The pressure of a fixed mass of gas is 200 kPa at 40°C and the volume is $1.5\,\text{m}^3$. The temperature is increased to 100°C but the volume remains the same. What is the new pressure?

13   The r.m.s. speed of a gas is $420\,\text{m s}^{-1}$ when the pressure is 100 kPa. If the volume remains the same, what pressure would the gas exert when the r.m.s. speed is $840\,\text{m s}^{-1}$?

14   A mass of $0.002\,\text{kg}$ of hydrogen contains 6×10^{23} molecules. Determine the number of molecules present in $0.5\,\text{m}^3$ of hydrogen at a pressure of 700 kPa and a temperature of 17°C. The density of hydrogen at s.t.p. is $0.09\,\text{kg m}^{-3}$.

15   A closed flask with a volume of 400 cm³ contains gas which has a mass of 5×10^{-4} kg. The temperature of the gas is 15°C.
What is the density of the gas?
The flask is cooled to 0°C. What can you say about
   i) the pressure of the gas
   ii) the density of the gas
   iii) the average kinetic energy of the molecules of the gas
compared to the values at 15°C?

16   A block of solid carbon dioxide measures 1 cm × 1 cm × 1 cm and has a mass of 0.0016 kg.
What is the density of the solid CO_2?
If the density of gaseous carbon dioxide is 1.98 kg m⁻³ at s.t.p., estimate the volume of gaseous carbon dioxide at s.t.p. obtained from this block.
How does the average spacing of the molecules in the solid CO_2 compare with that of the gas?

17   The initial pressure in container A is 130 kPa and the initial pressure in container B is zero.

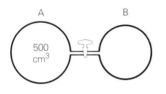

The tap is opened and the common pressure is 85 kPa. What is the volume of container B? Assume that the volume of the connecting pipe is negligible.

18   In a container with a volume of 2 m³, the pressure of a gas is 220 kPa. What mass of gas will be present if the r.m.s. speed of the molecules is 500 m s⁻¹?

19   The speeds of the molecules of a gas at s.t.p. are distributed as shown.

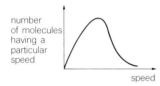

What difference would you expect in the distribution at 100°C?

20   The following equations refer to 1 mole of an ideal gas.
$pV = RT$ and $pV = \frac{1}{3}Nm\bar{c}^2$ where $N = 6 \times 10^{23}$ molecules

What are the units of R, the gas constant?
If 1 mole of any gas occupies a volume of 22.4×10^{-3} m³ at s.t.p., calculate a numerical value for R.
By combining these two equations, show that the average translational kinetic energy of a molecule is given by $\frac{3}{2}kT$ where $k = R/N$

21   The table shows the density and r.m.s. speeds of some gases at s.t.p.

gas	argon	oxygen	nitrogen	neon	helium	hydrogen
density (kg m⁻³)	1.80	1.43	1.25	0.90	0.18	0.09
r.m.s. speed (m s⁻¹)	410	461	493	581	1300	1840

Plot a graph of r.m.s. speed against density.
What is the relationship between r.m.s. speed and density?

22   In an agricultural project it is necessary to determine the average volume of a particular type of small seed. The apparatus below is devised to achieve this.

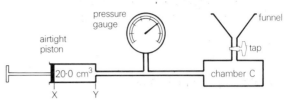

a) With the tap open and the piston at X, the pressure gauge indicates that atmospheric pressure is 1.02×10^5 Pa. The tap is closed and the piston is slowly pushed to Y. The gauge now reads 1.62×10^5 Pa. The internal volume between X and Y is 20.0 cm³.
Calculate the volume of air that is now in the apparatus, i.e. from Y into and including chamber C.
b) The tap is again opened and the piston returned to X. A sample of seeds is now poured through the funnel into C and the tap is closed. After the piston has been slowly pushed to Y, the pressure gauge reads 1.95×10^5 Pa.
   i) Calculate the new volume of air in the apparatus now.
   ii) What is the volume of the sample of seeds in C?
c) Explain briefly the effect (if any) on the accuracy of the measurement of the volume of the seed sample, if the volume inside the apparatus from Y into and including C was made very large compared with the volume between X and Y.
d) Describe how you would find the average volume of a single seed. (Counting all the seeds in the sample is not practicable.)
SCEEB

23   The kinetic theory of gases predicts that the product of the pressure (p) and the volume (V) of a fixed mass of gas is given by $pV = \frac{1}{3}Nm\bar{c}^2 = \frac{2}{3}N(\frac{1}{2}m\bar{c}^2)$
a) Use this expression to show that the root mean square speed of the molecules is given by $\sqrt{3p/\rho}$ where ρ is the density of the gas.
b) Calculate the root mean square speed of neon molecules at standard temperature and pressure. (Essential data can be found in the data book.)
c) Avogadro's hypothesis states that equal volumes of all gases contain the same number of molecules if the gases are at the same temperature and pressure. Write down the above expression for the product pV for each of two gases A and B and hence show how Avogadro's hypothesis can be deduced.
SCEEB

24   On a certain day the atmosphere pressure is 1.00×10^5 Pa (1 Pa = 1 N m⁻²) and the temperature is 7°C. A car pressure gauge which reads 0 Pa when open to the atmosphere reads 1.24×10^5 Pa when connected to a car tyre.
a) What is the reading on the gauge when it is connected to the tyre after the car has been standing in the sun until the temperature of the air in the tyre has risen to 27°C?
b) What is the ratio of the average (r.m.s.) speeds of the air molecules in the tyre at the two temperatures?
c) When asked an explanation of this effect in terms of kinetic theory, a pupil wrote:
   'If the temperature of a gas doubles, the speed of the molecules doubles and so does the momentum. Therefore the force exerted on the walls doubles and so does the pressure of the gas.'
This is not a satisfactory answer. Rewrite the explanation correctly.
SCEEB

25 The kinetic theory of gases predicts that the product of the pressure (p) and the volume (V) of a fixed mass of gas is given by the equation

$$p V = \tfrac{1}{3} N m \overline{c^2}$$

When experiments are carried out on a fixed mass of gas, the following conclusions can be drawn:

   i) The pressure is inversely proportional to the volume when the temperature is kept constant, ($p \propto 1/V$)
   ii) The pressure is directly proportional to the temperature when the volume is kept constant, ($p \propto T$)
   iii) The volume is directly proportional to the temperature when the pressure is kept constant, ($V \propto T$)

a) Choose one of the conclusions (i), (ii) or (iii) and describe how you would verify it in the laboratory. Include in your answer:

   a labelled sketch of the apparatus you would use;

   a description of how you would allow only **two** of the quantities p, V and T to change;

   a description of how you would make the measurements;

   and an indication of how you would use your results to arrive at the conclusion.

b) Show that each of the conclusions (i), (ii) or (iii) is consistent with the equation $p V = \tfrac{1}{3} N m \overline{c^2}$

SCEEB

26 a) It can be shown from theoretical considerations that where p is the pressure of a sample of gas, V is its volume, n is the number of molecules present, m is the mass of a molecule and $\overline{v^2}$ is the average of the squares of the velocities of the molecules, then

$$p V = \tfrac{1}{3} n m \overline{v^2} = \tfrac{2}{3} n \left(\tfrac{1}{2} m \overline{v^2} \right)$$

Write down the results which are obtained from three experiments, each of which is concerned with a different relation involving pressure, volume and temperature of a fixed mass of gas.
Show that each of these relations is consistent with the above equation.
b) Estimate the number of molecules of hydrogen contained in a cylinder of capacity 500 litres in which the gas is under a pressure of 10 atmospheres and is at a temperature of 15°C. (Take Avogadro's Number as 6×10^{23} molecules per gram mole.)
c) If the temperature of the gas in the cylinder in (*b*) is raised, what effect will this have on;
   i) the number of molecules in the cylinder;
   ii) the average velocity of the molecules;
   iii) the average distance between molecules;
   iv) the number of collisions per second between molecules?

SCEEB

27 a) The kinetic theory of gases predicts that the product of pressure (p) and volume (V) of a fixed mass of gas is given by the expression $p V = \tfrac{1}{3} N m \overline{c^2}$
   i) Discuss the significance of the term $\overline{c^2}$ in this expression.
   ii) State any four assumptions which are made in deriving the expression.
   iii) Show that the expression is consistent with the experimental result '$p V = $ constant', when the temperature of a fixed mass of gas is kept constant.
b) A substance has a density of 1000 kg m⁻³ in its liquid state. In its gas state, it has a density of approximately 1 kg m⁻³. Show that these figures are consistent with the average molecular separation in a gas being about ten times that in a liquid.

SCEEB

28 a) Describe an experiment to show how for a fixed mass of gas, kept at constant volume, the pressure varies with temperature. Your answer should include a sketch of the apparatus, a description of the experimental procedure and an indication of how you would use the results to find the relationship.
b) The kinetic theory of gases predicts that the pressure of a gas is given by the expression $p = \tfrac{1}{3} \rho \overline{c^2}$ where the symbols have their usual meaning.
Standard Temperature and Pressure (s.t.p.) is a temperature of 273 K and a pressure of 1.01×10^5 Pa.
Information on two gases, oxygen and hydrogen, at s.t.p. is given below:

   density of hydrogen $= 9.00 \times 10^{-2}$ kg m⁻³
   density of oxygen $= 1.43$ kg m⁻³

r.m.s. speed of oxygen molecules $= 4.60 \times 10^2$ m s⁻¹.
   i) Show that the r.m.s. speed of hydrogen molecules is approximately four times the r.m.s. speed of oxygen molecules at s.t.p.
   ii) To what temperature must the oxygen be raised so that its molecules have the same r.m.s. speed as the molecules of hydrogen at s.t.p.?

SCEEB

29 It is possible, using the following apparatus, to change the volume, temperature and pressure of a trapped mass of gas.

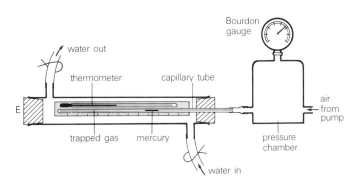

a) The temperature of the trapped gas is kept constant and the pressure increased. Sketch the shape of graph you would expect to obtain if the pressure of the gas were plotted against the length of the trapped gas column.
b) Water at 27°C surrounds the capillary tube and the gauge reads 1.00×10^5 N/m² (atmospheric pressure).
Air is pumped into the chamber until the gauge reads a steady value of 1.20×10^5 N/m². What is the new length l of the column of trapped gas if it was 26 cm originally?
c) The water temperature is now changed to 57°C.
   i) Should air be released from or pumped into the chamber to maintain the length of the column at the value l?
   ii) What will then be the reading on the Bourdon gauge?
d) The tube containing the trapped gas is now mounted vertically with end E uppermost. Explain whether the pressure of this gas as read from the gauge would be too high or too low.
e) With the aid of a diagram explain briefly how a Bourdon Gauge works and how it could be calibrated to read gas pressures in N/m².

SCEEB

14 The atomic model

It was at the start of the twentieth century that scientists began to build up a picture of the structure of the atom. A valuable contribution to this picture was made by R. A. Millikan in 1911 in his experiment to investigate the nature of negative charge.

14.1 Millikan's oil drop experiment

Millikan carried out experiments to measure the amount of charge held by small oil drops which had been charged by friction when they were sprayed through a narrow opening. In order to measure the charge on an oil drop, he carried out experiments to find the weight of each oil drop and the force exerted on the charged drop by an electric field.

If two parallel metal plates, a distance d metres apart, have a potential difference of V volts between them, the electric field strength between them is given by

$$E = \frac{V}{d}$$

The magnitude of an electric field is equal to the force per unit charge in that field. Thus if F_E is the force on a charge of q coulombs in an electric field of E volts per metre

$$E = \frac{F_E}{q}$$
$$\Rightarrow \quad F_E = E \times q$$

Millikan sprayed drops of oil, from a fine spray, into the space between two metal plates, Figure 14.1

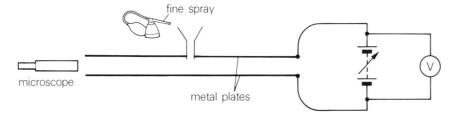

Figure 14.1 Millikan's oil drop apparatus

As the oil was forced through the fine spray, some of the drops became negatively charged by friction. Millikan used a microscope to observe the oil drops. There was a potential difference between the metal plates, the top plate being more positive, so that the electric field exerted an upward force on the negatively charged oil drops. Millikan adjusted the potential difference between the plates until an oil drop under observation became stationary. For the stationary oil drop, the net force acting on it was zero, since its acceleration was zero. This meant that the upward force exerted by the electric field on the oil drop was equal to the downward force exerted on it by the gravitational field.

Figure 14.2 shows the two forces acting on the stationary oil drop.

The upward force F_E exerted by the electric field is given by

$F_E = q \times E$ where E = electric field strength

q = charge on the oil drop

The downward force F_g exerted by the gravitational field is the weight of the oil drop and is given by

$F_g = mg$ where m = mass of the oil drop

g = gravitational field strength

Since the forces are equal for a stationary oil drop

$F_E = F_g$

$\Rightarrow qE = mg$

$E = \dfrac{V}{d}$ where V = potential difference between the plates

d = distance between the plates

$\Rightarrow \dfrac{qV}{d} = mg$

$\Rightarrow q = \dfrac{mgd}{V}$

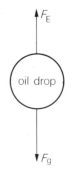

Figure 14.2 Forces exerted on a stationary oil drop

The gravitational field strength was known, and Millikan was able to measure the distance d and the potential difference V between the plates. In order to find q it was necessary to find the mass m of the oil drop.

With zero potential difference between the plates, there is no electric field and the oil drop falls. As the oil drop accelerates the frictional force increases and the small oil drop soon reaches its terminal velocity. This happens when the frictional force opposing the fall of the oil drop is equal and opposite to the weight of the drop, Figure 14.3.

The terminal velocity depends on the mass and volume of the oil drop and, if the terminal velocity is known, the mass can be calculated. Having measured the potential difference required between the plates to balance an oil drop, Millikan reduced the potential difference to zero and measured the terminal velocity of the drop by timing how long it took to fall through a measured distance. From these measurements he was able to calculate the mass of an oil drop. Knowing the mass of the oil drop m, the distance between the plates d and the potential difference between the plates V, he was able to calculate the charge q on the oil drop, using the equation.

$q = \dfrac{mgd}{V}$

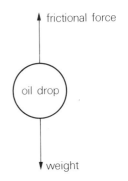

Figure 14.3 Forces on falling oil drop

When he had made the necessary measurements on an oil drop, Millikan changed the charge on the drop by irradiating the space between the metal plates with X-rays. The X-rays ionized air in the space, and the charged particles produced either increased the negative charge or removed negative charge from the oil drop. Millikan was then able to repeat the experiment to find the new charge on the drop.

Millikan repeated the experiment thousands of times and obtained a range of values for the charges on oil drops. From his results Millikan concluded that the charge values were all multiples of one basic value, -1.6×10^{-19}C, as illustrated in Table 1.

Millikan realized that the charges on the oil drops were all multiples of -1.6×10^{-19}C and concluded that this was the minimum quantity of charge that could exist. The different charges on the oil drops were due to different numbers of electrons and the charge on an electron is -1.6×10^{-19}C.

some values of q (C)

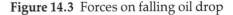

$-1.6 \times 10^{-19} = 1 \times (-1.6 \times 10^{-19})$

$-3.2 \times 10^{-19} = 2 \times (-1.6 \times 10^{-19})$

$-4.8 \times 10^{-19} = 3 \times (-1.6 \times 10^{-19})$

$-6.4 \times 10^{-19} = 4 \times (-1.6 \times 10^{-19})$

$-8.0 \times 10^{-19} = 5 \times (-1.6 \times 10^{-19})$

Table 1 Some results for q from Millikan's Experiment

14.2 Photoelectric effect

In Chapter 12 we discussed the fact that there were two conflicting theories of light, the Wave Theory and the Particle Theory. Thomas Young's experiments, producing interference patterns for light, provided strong evidence for the Wave Theory. Evidence supporting the Particle Theory was provided by the photoelectric effect.

In 1887 Heinrich Hertz found that when ultraviolet radiation shines onto two metal spheres, the potential difference required to produce a spark between them is reduced. The following year Hallwachs discovered that a negatively charged zinc plate lost its charge when it was exposed to ultraviolet radiation. In 1899 Lenard carried out further experiments which indicated that the ultraviolet radiation ejects electrons from some metals. These experiments were all demonstrations of what is known as 'the photoelectric effect'.

This effect can be demonstrated by a simple laboratory experiment, Figure 14.4. An electroscope with a polished zinc plate on its disc is negatively charged. When ultraviolet radiation is shone onto the zinc plate the electroscope rapidly discharges.

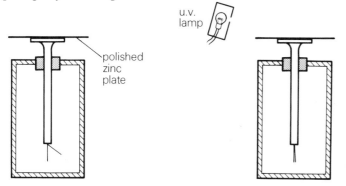

Figure 14.4 Photoelectric effect

At first this may not appear to be very surprising. Electroscopes are discharged by simply bringing a flame near to them. This happens because the energy of the flame ionizes some of the molecules in the surrounding air, forming negative and positive ions. A negatively charged electroscope attracts the positive ions which collect electrons from the electroscope and cause it to discharge, Figure 14.5.

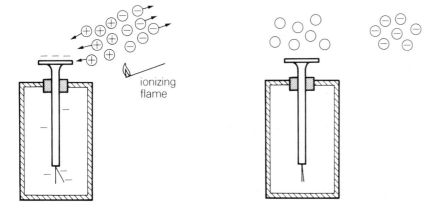

Figure 14.5 Discharge of a negatively charged electroscope

If the electroscope is initially positively charged, it is the negative ions which are attracted to it and cancel out its charge, Figure 14.6.

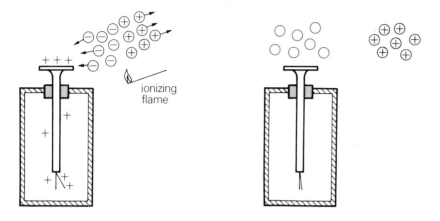

Figure 14.6 Discharge of a positively charged electroscope

Other ionizing sources such as alpha particles or X-rays discharge electroscopes in the same way. Ultraviolet radiation discharges a negatively charged electroscope with a zinc plate but, if the same electroscope is positively charged, the ultraviolet radiation does not discharge it, Figure 14.7.

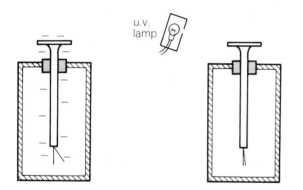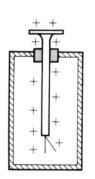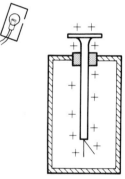

Figure 14.7 Effect of ultraviolet radiation on charged electroscopes

This shows that the ultraviolet radiation does not discharge the electroscope by ionizing the air.

Ultraviolet radiation discharges the negatively charged zinc plate of an electroscope. Light, which is an electromagnetic radiation of a lower frequency than ultraviolet, does not discharge such an electroscope. It seems that ultraviolet radiation has sufficient energy to eject electrons from the zinc, while visible radiation does not. If the case were that simple, increasing the brightness of the light would increase the total energy supplied and, if the light were bright enough, the electroscope would be discharged. This is not the case. No matter how bright the visible radiation, the electroscope is not discharged. However, a relatively small amount of ultraviolet radiation will cause it to discharge.

The explanation of the photoelectric effect was that electromagnetic radiation consists of small particles or corpuscles.

The radiation will eject electrons only if the corpuscles have sufficient energy. The energy of a corpuscle of visible radiation is less than the energy required to remove an electron from zinc and, no matter how many corpuscles of visible radiation are supplied, no electrons are ejected. The energy of a corpuscle of ultraviolet radiation is sufficient to eject an electron from zinc.

The idea that radiation consists of corpuscles rather than continuous waves is support for the corpuscular theory. The simple demonstration with the gold leaf electroscope provides evidence for this theory. The evidence can be summarized as follows:

evidence	conclusion
1 Ultraviolet radiation discharges the zinc plate of an electroscope which is negatively charged but not of one which is positively charged.	Discharge is a result of ejecting electrons and not a result of ionizing the air about the electroscope.
2 Visible radiation, no matter how bright, does not produce the same effect.	It is *not* simply a case of the total energy supplied, but rather a case of whether each 'bundle' of radiation has sufficient energy to eject an electron.

Young's slits experiment provided strong evidence for the wave theory for electromagnetic radiation. The photoelectric effect provided strong evidence for the corpuscular theory. Einstein provided an explanation which embodied both theories. He suggested that the corpuscles could be thought of as bundles of wave energy, called photons, Figure 14.8.

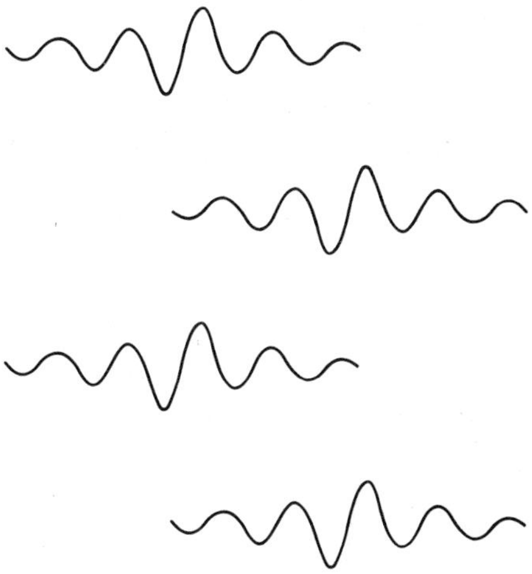

The energy E of each photon is proportional to the frequency f of the radiation.

$$E \propto f$$
$$\Rightarrow \quad E = h \times f \qquad \text{where } h \text{ is a constant}$$

The constant h is known as **Planck's constant** and was named after Max Planck who first suggested that the frequency of radiation emitted by an atom was directly proportional to the amount of energy radiated by the atom. The value of h is 6.6×10^{-34} J s.

Figure 14.8 Photons of wave energy

Example 1

The energy required to eject an electron from sodium is 2.9×10^{-19} J. What is the minimum frequency of electromagnetic radiation required to produce the photoelectric effect with sodium?

energy required to eject an electron $= 2.9 \times 10^{-19}$ J

energy of a photon of radiation $\quad = h \times f$

where f = frequency of the radiation
and h = Planck's constant

For the photoelectric effect to occur with sodium

$$h \times f = 2.9 \times 10^{-19}$$
$$\Rightarrow \quad f = \frac{2.9 \times 10^{-19}}{h}$$
$$h = 6.6 \times 10^{-34}$$
$$\Rightarrow \quad f = \frac{2.9 \times 10^{-19}}{6.6 \times 10^{-34}} = 4.4 \times 10^{14} \text{ Hz}$$

Minimum frequency $= 4.4 \times 10^{14}$ Hz

In 1923, R. A. Millikan who is best known for his oil drop experiment, won the Nobel prize for his work on the photoelectric effect. He carried out many careful experiments, finding the minimum frequency of radiation required to produce the photoelectric effect in different metals.

Figure 14.9 shows a typical apparatus for finding the minimum frequency of radiation to emit electrons from metals and measuring the rate at which electrons are emitted.

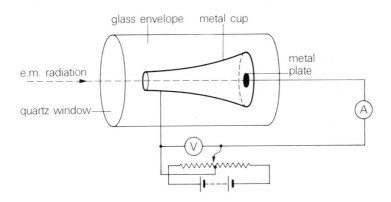

Figure 14.9 Photoelectric effect apparatus

The electromagnetic radiation passes through a hole in the metal cup onto a metal plate. This plate is made of the metal for which the photoelectric effect is being investigated. Any electrons emitted from the metal plate may cross to the metal cup or return to the metal plate, depending on the potential difference between the plate and the cup. A quartz window is used at the end of the glass tube because quartz allows both light and ultraviolet radiation to pass through it while glass absorbs ultraviolet radiation.

When both the frequency and the intensity of the electromagnetic radiation remain constant and the potential difference between the cup and the plate is varied, the current varies as shown by the graph in Figure 14.10.

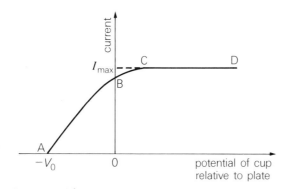

Figure 14.10 Variation of photoelectric current with potential difference

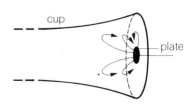

Figure 14.11 Reverse potential stops current

On the graph, negative values of potential difference represent potential differences which oppose the flow of electrons from the metal plate to the metal cup. The minimum value V_0 of the reverse potential which reduces the photoelectric current to zero, is called the stopping potential, Figure 14.11.

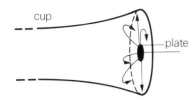

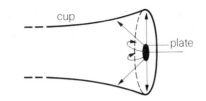

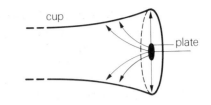

Figure14.12 Some of the electrons overcome the reverse potential

Figure 14.13 Potential supports the flow of electrons

Figure 14.14 All of the photoelectrons reach the plate

For the section AB on the graph, the potential difference between the plate and the cup opposes the flow of electrons from the plate to the cup, but is insufficient to overcome the kinetic energy of all the electrons. Thus the current in this section consists of the flow of electrons which have sufficient kinetic energy to overcome the opposing potential difference, Figure 14.12.

For the section BC on the graph, the potential difference supports the electron flow from plate to cup, but not all of the electrons reach the cup, Figure 14.13.

For the section CD on the graph, increases in the potential difference supporting the electron flow produced no increase in current. This is because all of the photoelectrons reach the plate and an increase in the potential difference does not increase the current because there are no more electrons available, Figure 14.14.

Example 2

If the maximum photoelectric current in the apparatus shown in Figure 14.9 is $50\,\mu A$, how many electrons are emitted per second?

current $= 50 \times 10^{-6}\,A$

charge flow per second $= 50 \times 10^{-6}\,C$

charge on an electron $= 1.6 \times 10^{-19}\,C$

$\Rightarrow$ number of electrons emitted per second $= \dfrac{50 \times 10^{-6}}{1.6 \times 10^{-19}}$

number of electrons emitted per second $= 3.1 \times 10^{14}$

Number of electrons per second $= 3.1 \times 10^{14}$

If the experiment is repeated several times with different intensities of electromagnetic radiation (all with the same frequency), the results produce a set of curves as shown in Figure 14.15.

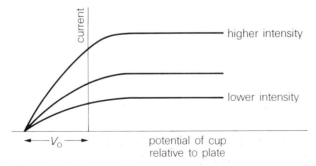

Figure 14.15 Photoelectric current for different intensities of radiation

These results show that the photoelectric current, which is equal to the maximum current in the circuit, depends on the intensity of the radiation. In fact, the photoelectric current is directly proportional to the intensity of the radiation. Increasing the intensity increases the number of photons per second and hence the number of electrons ejected per second.

The results also show that the stopping potential V_0 is independent of the intensity of the radiation; V_0 is the minimum potential difference required to stop all, including the most energetic, of the electrons emitted from the plate. Thus the fact that V_0 is independent of the intensity of radiation indicates that the energy of the electrons emitted is independent of the intensity of the radiation.

When a charge of Q coulombs moves through a potential difference of V volts, the work done on the charge is QV joules. Thus, if V_0 is the potential difference required to overcome the kinetic energy of the electrons which are emitted with most energy, the work done in stopping those electrons is qV_0 (q = charge on an electron). The work done on the electron is equal to the kinetic energy lost by that electron. Thus the maximum kinetic energy of an emitted electron is given by

$$E_{max} = qV_0 \qquad \text{where } V_0 = \text{ stopping potential}$$
$$q = \text{ charge on an electron}$$

If the experiment is repeated using electromagnetic radiations of different frequencies and V_0 is measured in each case, the results obtained are as illustrated in Figure 14.16.

The frequency f_0 is the minimum frequency for which the photoelectric effect occurs and is known as the cut-off frequency for the metal being used. The results shown in Figure 14.16 are those obtained by Millikan for sodium; in that case he found the cut-off frequency to be 4.39×10^{14} Hz. For radiations of frequency less than f_0, the individual photons do not have sufficient energy to eject electrons from the metal. Thus the minimum energy required to eject an electron from a metal is given by

$$E_{min} = hf_0 \qquad \text{where } h = \text{ Planck's constant}$$
$$f_0 = \text{ cut-off frequency for that metal}$$

From the experiments carried out with this apparatus we know that

1   the minimum frequency of radiation required to eject an electron from a metal is given by

$$E_{min} = hf_0 \qquad \text{where } f_0 = \text{ minimum cut-off frequency;}$$

2   the maximum kinetic energy of an ejected electron is given by

$$E_{max} = qV_0 \qquad \text{where } V_0 = \text{ stopping potential}$$
$$q = \text{ charge on an electron.}$$

These two energies may be related to the energy of the photons of radiation producing the photoelectric effect since, when a photon is absorbed, its energy ejects the electron with a certain amount of kinetic energy.

$$\frac{\text{energy of}}{\text{1 photon}} = \frac{\text{energy needed to}}{\text{eject electron}} + \frac{\text{kinetic energy given}}{\text{to that electron}}$$

The electron which needs least energy to be ejected will gain the most kinetic energy.

$$\frac{\text{energy of}}{\text{1 photon}} = \frac{\text{minimum energy required}}{\text{to eject an electron}} + \frac{\text{maximum kinetic energy of}}{\text{an emitted electron}}$$

If the frequency of the radiation producing the effect is f,

$$hf = hf_0 + qV_0$$

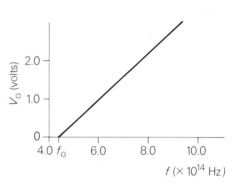

Figure 14.16 Variation of stopping potential with frequency of radiation

Example 3

Sodium has a cut-off frequency of 4.4×10^{14} Hz. What is the stopping potential when the sodium is irradiated with light of frequency 6.0×10^{14} Hz?

$$f = 6.0 \times 10^{14} \text{ Hz}$$

$$f_0 = 4.4 \times 10^{14} \text{ Hz}$$

$$h = 6.6 \times 10^{-34} \text{ J s}$$

$$q = 1.6 \times 10^{-19} \text{ C}$$

$$hf = hf_0 + qV_0$$

$$\Rightarrow 6.6 \times 10^{-34} \times 6.0 \times 10^{14} = 6.6 \times 10^{-34} \times 4.4 \times 10^{14} + 1.6 \times 10^{-19} \times V_0$$

$$\Rightarrow \qquad V_0 = \frac{6.6 \times 10^{-34}(6.0 - 4.4) \times 10^{14}}{1.6 \times 10^{-19}}$$

$$\Rightarrow \qquad V_0 = \frac{6.6 \times 1.6 \times 10^{-20}}{1.6 \times 10^{-19}}$$

$$\Rightarrow \qquad V_0 = 0.66$$

Stopping potential = 0.66 V

14.3   Spectra

The photoelectric effect provided evidence for the discrete rather than the continuous nature of electromagnetic radiation. This means that the radiation consists of 'bundles' of energy called photons. In the photoelectric effect, photons are absorbed and give energy to electrons in the atoms. The reverse also occurs. This is when electrons in the atoms lose energy and that energy is given off as electromagnetic radiation. The radiation given off when this occurs can be dispersed by a prism or diffracted by a diffraction grating. The spectrum formed in this way is different from that formed from the light from a filament lamp or the sun.

Emission spectra

When light is given off from a light source, it is split into its different colours by a prism or diffraction grating and forms a spectrum; such a spectrum is called an emission spectrum. Emission spectra are found to be of two types, continuous spectra and line spectra, Figure 14.17. Examples of these in colour are given on the back cover of the book.

Continuous spectra are produced by light from sources which are solids, liquids or high-pressure gases raised to high temperatures. Common examples of such sources are lamp filaments and the Sun. Line spectra are produced by light from gas discharge tubes, such as those used in neon lights and sodium street lights, or from hot gases and vapours.

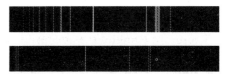

Figure 14.17 Emission spectra

Line emission spectra

A line emission spectrum consists of narrow lines of colour, showing that only radiations of specific frequencies are emitted.

The electric current through the gas gives some energy to the atoms in the gas and this energy gives electrons in the atoms extra energy. When this happens the electrons are said to be 'excited' to higher energy levels.

The atoms which have gained energy in this way then tend to return to their original more stable state, giving off their surplus energy. It is this surplus energy that is the radiation given off. In any such light source their are millions of atoms absorbing and emitting energy, yet the emission spectrum consists of a limited number of wavelengths. The reason for this is that, for atoms of a particular element, the electrons can have only a limited number of energy values, known as 'energy levels'. Figure 14.18 illustrates how an electron which has been excited to a higher energy level E_3 may return to its more stable energy level E_0, either in one step or by a number of steps through other permitted energy levels, E_1 and E_2.

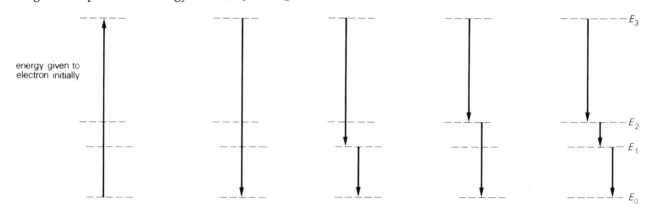

energy given to electron initially

Figure 14.18 Changes in energy levels

The most stable level E_0 is known as the 'ground level'. In the example shown in Figure 14.18, there are four different ways in which the electron can descend from energy level E_3 to the ground level. When an electron descends from one energy level to another, it loses energy and this energy is emitted as a photon of radiation. The energy E of a photon is given by

$E = hf$ where h = Planck's constant

         f = frequency of the radiation.

Thus, if the electron descends from energy level E_x to E_y, the change in energy E is given by

$E = E_x - E_y$

energy of the photon emitted = hf_{xy} where f_{xy} = frequency of the photon

energy of the photon emitted = decrease in the energy of the electron

$hf_{xy} = E_x - E_y$

$\Rightarrow f_{xy} = \dfrac{E_x - E_y}{h}$

Figure 14.19 shows the four possible ways in which the electron can descend from level E_3 to E_0 and the frequency of each photon emitted (the higher the wavelength the smaller the frequency).

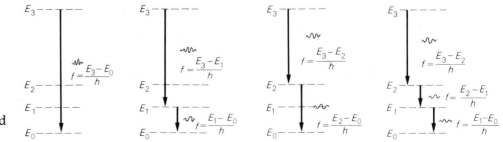

Figure 14.19 Photons of emitted radiation

An electron may descend by one of these ways only, but, with millions of atoms in the light source, electrons in different atoms will descend in different ways and the light emitted will contain the full range of frequencies shown in Figure 14.19. There are six different frequencies in this example. Not all of the frequencies are necessarily in the visible range of the electromagnetic spectrum.

In a line spectrum, the frequency of each radiation corresponds to the frequency of photons emitted by electrons descending between two definite energy levels.

Study of line spectra shows that some lines are brighter than others. This is because the electrons are more likely to occupy some energy levels than others. If in the example illustrated in Figure 14.19, E_1 is an energy level which is less likely to be occupied, the number of atoms in which the electrons descend by steps involving level E_1 will be less. This means that, in the spectrum, the lines produced by these steps would be less bright because there would be fewer photons of the energy corresponding to these steps.

In the example given, the frequencies of the lines that would be less bright if level E_1 is less likely to be occupied are:

$$\frac{E_3 - E_1}{h} \quad ; \quad \frac{E_1 - E_0}{h} \quad ; \quad \frac{E_2 - E_1}{h}$$

The fact that line spectra are produced by radiations from millions of atoms in the source shows that the permitted energy levels are the same in the many atoms of the element. In fact, these permitted energy levels are a characteristic of an element, and can be used to identify the element. Chemists use this fact to analyse very small samples. The sample is excited by an electric spark or simply by heating in a flame, and the light given off is dispersed to form a spectrum which is photographed. The photograph is compared with those of the spectra of different elements to see which elements are present in the sample.

By studying the spectra produced from the light emitted by stars, astronomers can determine which elements are present in the stars.

When an electron is not attached to an atom, it is said to be at an energy level of zero. When the electron becomes attached to the atom, energy is given off and this means that the energy of the electron is reduced below zero and is, therefore, negative. Since energy levels in an atom are the possible electron energies, energy levels must have negative values.

Example 4

The diagram represents three possible energy levels of an atom of hydrogen.

When the energy of an electron changes from a higher to a lower energy level, a quantum of electromagnetic radiation is emitted. The frequency f of the radiation emitted in a transition between energy levels E_1 and E_2 is given by

$$E_2 - E_1 = hf \quad \text{where } h \text{ is Planck's constant.}$$

a) How many lines in the hydrogen spectrum are produced as a result of transitions between the energy levels shown in the diagram?

b) Calculate the wavelength of one of these hydrogen spectrum lines.

$$h = 6.63 \times 10^{-34}\,\text{J s}$$

energy

-2.4×10^{-19} J

-5.4×10^{-19} J

-21.8×10^{-19} J

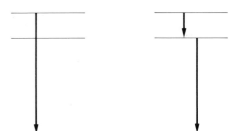

a) The possible ways that electrons might descend from the highest energy level to the lowest are shown in the diagram.

This shows that there are three possible changes in energy levels and this means that these changes will produce three lines in the hydrogen spectrum.

b) For the change from the highest energy level to the lowest,

$$E_2 = -2.4 \times 10^{-19} \, \text{J}$$
$$E_1 = -21.8 \times 10^{-19} \, \text{J}$$
$$h = 6.63 \times 10^{-34} \, \text{J s}$$
$$E_2 - E_1 = hf \qquad \text{where } f = \text{frequency of emitted photon}$$
$$\Rightarrow \quad -2.4 \times 10^{-19} - (-21.8 \times 10^{-19}) = 6.63 \times 10^{-34} \times f$$
$$\Rightarrow \quad f = \frac{19.4 \times 10^{-19}}{6.63 \times 10^{-34}}$$
$$\Rightarrow \quad f = 2.93 \times 10^{15}$$
$$v = f \times \lambda$$

For electromagnetic radiation in air $v = 3.0 \times 10^8 \, \text{m s}^{-1}$

$$3 \times 10^8 = 2.93 \times 10^{15} \times \lambda$$
$$\Rightarrow \quad \lambda = \frac{3 \times 10^8}{2.93 \times 10^{15}}$$
$$\Rightarrow \quad \lambda = 1.02 \times 10^{-7} \text{m}$$

Wavelength of emitted radiation $= 1.02 \times 10^{-7}$ m

Continuous emission spectra

In a solid, liquid or high-pressure gas, the atoms are much closer than in a normal gas. Under these circumstances, some of the outer electrons in the atoms experience forces from the neighbouring atoms as well as from the nucleus of the atom containing the electrons. At high temperatures the high energy of the atoms combined with these interatomic forces result in some of the electrons being able to take on a wide range of energies rather than being confined to the small number of energy levels within one atom. Since the electrons may now occupy a wide range of energy values, and may therefore undergo a wide range of energy changes, the radiation emitted contains a range of frequencies. In this way, light emitted by hot sources of radiation in which the atoms are relatively close together, contains the range of frequencies that produces a continuous spectrum.

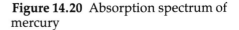

Figure 14.20 Absorption spectrum of mercury

Absorption spectra

White light emitted by a hot source produces a continuous spectrum when viewed through a spectroscope. If the white light is passed through a gas before entering the spectroscope, some dark lines are seen in the spectrum, Figure 14.20. The dark lines show that the gas absorbs some of the light, but only at certain definite frequencies. The spectrum formed in this way is called the absorption spectrum of the gas which absorbed the light.

The frequencies of absorbed light are identical to those in the line emission spectrum of the same gas, Figure 14.21.

The absorption spectrum is a result of photons of energy being absorbed by the atoms and exciting the electrons to higher energy levels. Only those photons with exactly the energy required to excite the electrons are absorbed.

Figure 14.21 Emission and absorption spectra of sodium

Thus the process that is occurring is the reverse of that described for the production of a line emission spectrum, Figure 14.22.

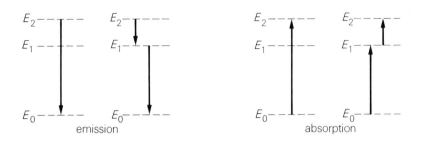

Figure 14.22 Transitions for line emission and absorption spectra

When an atom has absorbed a photon, it is in the excited state and is likely to return to the stable state by re-emitting the photon. At first sight it might seem that, since the photons are re-emitted, there should be no dark lines and hence no absorption spectrum; yet experiment shows this not to be true. The reason for this is that the light beam from which the light is absorbed is directed towards the spectroscope. When the light is re-emitted it is emitted in all directions, and only a small percentage of it will reach the spectroscope.

Absorption spectra can be used to analyse a gas through which light has passed.

Dark lines are observed in the spectrum of light produced from the Sun's rays. These are due to absorption by gases in the outer atmosphere of the Sun and study of these spectra has enabled astronomers to analyse the composition of this outer atmosphere. Similarly, absorption spectra of light which has passed through the atmospheres of some of the planets have produced evidence of the compositions of these atmospheres.

Summary

In Millikan's Oil Drop Experiment the charge q on a balanced oil drop is given by:

$q = \dfrac{mgd}{V}$ where d = distance between the plates

V = potential difference between the plates

m = mass of the oil drop

g = gravitational field strength

The photoelectric effect is demonstrated by the fact that ultraviolet light discharges a negatively charged zinc plate on an electroscope.

Electromagnetic radiation is made up of photons of wave energy and the energy E of a single photon of light is given by:

$E = h \times f$ where h = Planck's constant

f = frequency of the radiation.

In the photoelectric effect the maximum kinetic energy of an emitted electron is given by:

$E_{max} = qV_0$ where V_0 = stopping potential

q = charge on an electron.

In the photoelectric effect, the minimum frequency of radiation required to eject an electron from the metal is given by:

$E_{min} = h \times f_0$ where f_0 = cut-off frequency

h = Planck's constant.

A line emission spectrum consists of lines of specific frequencies of light. Each frequency corresponds to the energy of a photon equal to the energy emitted by an electron descending from one permitted energy level to another.

For an electron descending from energy level E_x to E_y the frequency f_{xy} of the emitted photon is given by:

$f_{xy} = \dfrac{E_x - E_y}{h}$

An absorption spectrum is formed when white light is passed through a gas or vapour and lines of specific frequencies of light are absorbed. Each frequency absorbed corresponds to the energy of a photon equal to the energy of an electron ascending from one permitted energy level to another.

Problems

1 An experimenter carried out Millikan's Oil Drop experiment and obtained the following results for the charges on an oil drop:-

1.6×10^{-19} C; 0.8×10^{-19} C; 0.8×10^{-19} C; 1.2×10^{-19} C; 2.0×10^{-19} C; 1.6×10^{-19} C.

From these values what should he deduce as the value of the charge on an electron? This value does not agree with the accepted value for the charge on an electron and the error was due to the fact that the experimenter had misread the range on the voltmeter he was using. He thought that the voltmeter was on a setting to read from 0 to 10 000 V. By comparing his value for the charge on an electron with the accepted value, deduce the actual range of the voltmeter that he was using.

2 Two metal plates, 0.5 cm apart, have a potential difference of 1000 V between them.
a) What is the magnitude of the electric field between the plates?
b) An oil drop of mass 6.4×10^{-15} kg is stationary between the plates. What is the charge on the oil drop?

3 An experimenter investigated the effect of shining different electromagnetic radiations on the polished zinc plates of charged electroscopes. He found that ultraviolet radiation discharged the negatively charged electroscope but not the positively charged electroscope. Visible radiation, no matter how bright, did not discharge any electroscope.
a) How does this experimental evidence support the following statements?
　i) The process by which the ultraviolet radiation discharges the electroscope is not the ionization of the surrounding air.
　ii) Whether or not the electroscope is discharged does not depend on the total amount of energy supplied by the electromagnetic radiation.
b) What is the name of the 'effect' by which ultraviolet radiation can eject electrons from zinc?
c) For which theory did this effect provide strong support?

4 For a certain metal, the energy required to eject an electron from an atom is 3.3×10^{-19} J.
a) What is the minimum frequency of electromagnetic radiation required to produce the photoelectric effect with this metal?
b) Would the photoelectric effect occur when this metal is illuminated with light of:
　i) frequency 4×10^{14} Hz; ii) wavelength 5×10^{-7} m?

(Planck's constant = 6.63×10^{-34} J s)

5 Ultraviolet radiation of frequency 3.0×10^{16} Hz ejects an electron from an atom of an element. If the electron leaves the atom with a kinetic energy of 5×10^{-18} J, calculate:
a) the minimum amount of energy that was necessary to eject the electron from the atom;
b) the minimum frequency of radiation that is necessary to produce the photoelectric effect with this element.

(Planck's constant = 6.63×10^{-34} J s)

6 In an experiment to investigate the photoelectric effect for a certain metal, the cut-off frequency is found to be 6.0×10^{16} Hz. What is the stopping potential when the metal is irradiated with radiation of frequency 6.4×10^{16} Hz?

(Planck's constant = 6.63×10^{-34} J s;
charge on electron = 1.6×10^{-19} C)

7 The diagram represents four possible energy levels of an atom of a metal.

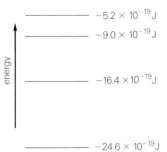

-5.2×10^{-19} J

-9.0×10^{-19} J

-16.4×10^{-19} J

-24.6×10^{-19} J

a) How many lines in the spectrum of this metal are produced as the result of transitions between the energy levels shown in the diagram?
b) Calculate the wavelengths of the spectrum lines representing the greatest and the least energy transitions.
(Planck's constant $= 6.63 \times 10^{-34}$ J s)

8 How does the line emission spectrum of an element compare with its absorption spectrum?

9 A beam of ultraviolet radiation falls on a suitable metal plate Y, which lies on the axis of a hollow metal cylinder X. X and Y are connected in an electric circuit including a battery and a milliammeter as shown in the diagram.

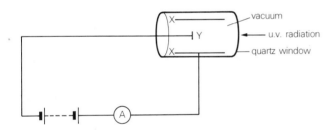

a) Explain why a small current is registered.
b) What would happen to the current if the intensity of the light were increased?
SCEEB

10 The diagram shows part of the emission spectrum of an element.

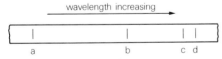

Light of frequency corresponding to each of the above spectral lines is allowed to strike a metal plate in turn and in some cases electrons are ejected from the metal.
i) Light from which of the above spectral lines is most likely to eject electrons from the plate? Give a reason for your answer.
ii) Light of frequency 5.08×10^{14} Hz, corresponding to one of the above lines, can eject electrons with a kinetic energy of 0.45×10^{-19} J from the metal plate. How much energy is required just to release electrons from the metal?
iii) Show whether light of frequency 4.29×10^{14} Hz, corresponding to line c, is capable of ejecting electrons from the metal.
(Planck's constant $h = 6.63 \times 10^{-34}$ J s)
SCEEB

11 In Planck's Quantum Theory of Light, the energy E of a quantum of light (photon) is given by the equation $E = hf$ where f is its frequency and h is Planck's constant, the value of which is 6.63×10^{-34} J s. The minimum energy required to eject an electron from a certain metal is 3.00×10^{-19} J. Explain whether you would expect light of wavelength 5.00×10^{-7} m to eject electrons from the metal. What is the name given to this phenomenon?
SCEEB

12 Certain metals are observed to emit electrons when irradiated with ultraviolet light. (The photoelectric effect.)

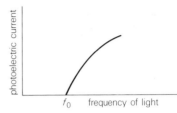

a) The graph indicates that there is no photoelectric current if the frequency of the light is below a certain value f_0. How can this be explained?
b) Discuss **briefly** the consequences of the interpretation of the photoelectric effect on the theory of the nature of light held at the time of its discovery.
SCEEB

13 a) Explain what is meant by the photoelectric effect. Indicate how it depends on
i) the frequency of the light;
ii) the intensity of the light.
Explain how your answers to (i) and (ii) are related to a theory of the nature of light.
b) A clean zinc plate is mounted in an ionization chamber, just below a wire mesh as shown below. The chamber is connected in series with a d.c. supply and a sensitive current meter. The current meter amplifies any small current in the circuit by a factor of 10^6 and displays the amplified current on a microammeter. The zinc plate is illuminated by a ultraviolet lamp.

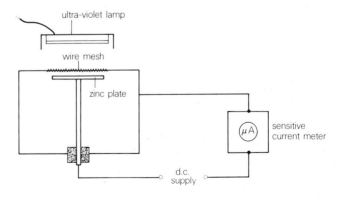

Describe how you would use the apparatus to show that any small current in the circuit was due to the photoelectric effect.
SCEEB

15 A nuclear model

15.1 Introduction

At the beginning of the nineteenth century John Dalton put forward a theory which assumed that matter consisted of solid indivisible particles called atoms. At the start of the twentieth century, experimental evidence had suggested the existence of even smaller particles. It was recognized that atoms contain positive and negative charges; J. J. Thomson suggested a model in which the positive charge is distributed evenly throughout the volume of the atom with negative charges fixed at various points, like 'currants in a plum pudding'. On the basis of this model, a stream of charged particles fired at the atom would be deflected through only fairly small angles.

In 1908 Geiger and Marsden, two assistants of Ernest Rutherford, started to investigate the deflection of alpha particles which they fired at a thin metal foil, Figure 15.2.

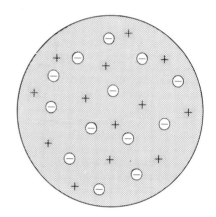

Figure 15.1 Model of the atom suggested by J. J. Thomson

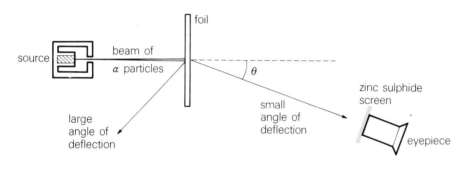

Figure 15.2

In 1909 they made the surprising discovery that a few of the particles bounced back from the foil. When Geiger reported to Rutherford, he expressed his surprise by saying that 'it was almost as incredible as if you had fired a 15-inch shell at a piece of tissue paper and it came back and hit you'.

Rutherford showed that these results could only be explained if the positive charge was concentrated into a very small volume. Making this assumption Rutherford applied the rules of electrostatic force between charges and predicted the number of particles which would be deflected along a given direction.

Geiger and Marsden then tested Rutherford's theory using the apparatus shown in Figure 15.3.

In the experiments, a beam of alpha particles from a radon source was fired at a metal foil. After deflection the particles were observed through the eyepiece which was set at different positions round the rotating table. The presence of the alpha particles was detected by a zinc sulphide screen. Every time an alpha particle hit the screen, a minute flash of light (a scintillation) was seen through the microscope. By counting the number of flashes, it was possible to determine the number of particles arriving at the screen. The screen could be moved to detect alpha particles which had been deflected at various angles. The chamber was evacuated to prevent the absorption of the particles by the air.

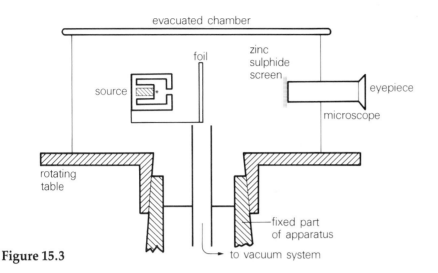

Figure 15.3

In his theory Rutherford predicted four factors upon which the number of particles hitting the zinc sulphide screen would depend.

1 the magnitude of the positive charge on the nucleus

2 the angle of deflection θ

3 the velocity of the particles

4 the thickness of the metal foil.

The experimental results confirmed that these four factors determined the number of particles hitting the screen. On the basis of these findings Rutherford proposed a model of the atom in which the mass and positive charge are concentrated in the nucleus surrounded by a space containing the negative charge.

To appreciate the relative sizes, if the diameter of the nucleus were a few millimetres the whole of the atom would be about the same size as a football stadium. It is not possible to draw this to scale but Figure 15.4 gives a simple picture of the model.

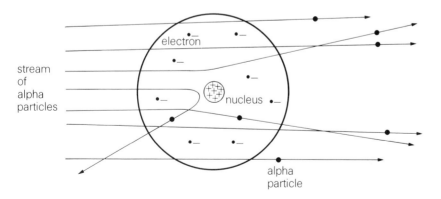

Figure 15.4

The particles in the nucleus were called protons by Rutherford (after the Greek *protos* meaning 'first'). This was taken up by Niels Bohr (among others) who proposed that the negative charges, the electrons, were confined to strictly defined orbits round the nucleus. The charge on an electron is -1.6×10^{-19}C and the charge on a proton is $+1.6 \times 10^{-19}$C. Since the atom is electrically neutral, the total negative charge is equal to the total positive charge. Thus, for a neutral atom, the total number of electrons must be the same as the total number of positive charges. All the atoms of any one element contain the same number of protons; this number is called the atomic number.

15.2 Isotopes

A mass spectrometer is used to determine accurately the mass of the atoms of an element. A simplified diagram of the apparatus is shown in Figure 15.5.

Positive ions of the element under test are produced at an anode and accelerated through a hollow cathode. They then pass through an electric field which produces a deflection away from the positive plate (plate A in Figure 15.5).

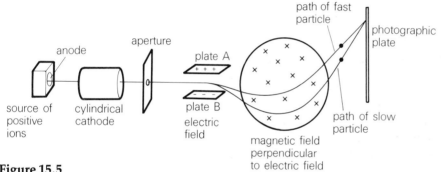

Figure 15.5

The positive ions are then deflected by a magnetic field which causes the particles to hit a photographic plate and form an image. By using these two fields, it is possible to eliminate the effects of varying speeds in the particle stream so that all particles of the same mass are focused at one point only on the screen. It is possible to determine accurately the mass of any particle by measurements taken from the mark on the photographic plate.

Data obtained on the mass of atoms indicated that atoms of the same element can have different masses. Such atoms are called **isotopes**. It is possible for an atom to have more than one isotope; for example, the element tin has as many as ten isotopes. An explanation for the existence of isotopes was finally provided in 1932 by James Chadwick. He bombarded a beryllium target with alpha particles, Figure 15.6.

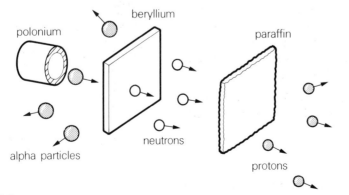

Figure 15.6

The beryllium target gave off radiation which was found to be able to penetrate several centimetres of lead. If however, this radiation is used to bombard paraffin wax the radiation is absorbed and protons are ejected from the slab. By considering conservation of momentum and energy, Chadwick was able to show that a neutral particle was emitted by the beryllium. He called this particle a neutron, symbol $_0^1$n, which has the same mass as a proton but no charge. It is the presence of the neutron in the nucleus which accounts for isotopes. Isotopes of an element have the same number of protons but different numbers of neutrons.

When specifying an atomic nucleus it is necessary to give two numbers

Z the atomic number   = number of protons in nucleus

A the mass number   = number of protons + number of neutrons

For example a beryllium nucleus has the symbol $_4^9$Be where the number of protons is 4 and the number of neutrons is 5.

These symbols are used in an equation to describe a nuclear reaction. For example, when an alpha particle bombards a beryllium nucleus, a neutron and a carbon nucleus are formed.

$$_4^9\text{Be} \quad + \quad _2^4\text{He} \quad \longrightarrow \quad _6^{12}\text{C} \quad + \quad _0^1\text{n}$$

beryllium   alpha particle   carbon   neutron

It is known that atoms are made up of protons, neutrons and electrons. The protons and neutrons are in the nucleus and are collectively known as nucleons. A nucleus of any specific element $_Z^A$X is termed a nuclide. For nuclei of greater mass, greater numbers of nucleons are packed together. For example, the hydrogen nucleus $_1^1$H contains 1 proton and no neutrons, but a uranium nucleus $_{92}^{238}$U contains 92 protons and 146 neutrons. Since protons are positive, they exert an electrostatic force of repulsion on each other, yet are contained tightly packed together in the nucleus without flying apart. Part of the reason for this lies in the presence of the neutrons. Investigations show that an element having an even number of protons in the nucleus has more isotopes than one with an odd number of protons, while no element with an odd atomic number has more than two stable isotopes. Often only one isotope of an element exists, for example, gold $_{79}^{197}$Au.

15.3   Binding energy

Neutrons play a part in holding protons together within the nucleus. The force which holds the nucleons together is not completely understood. It appears to be some kind of exchange force which binds protons and neutrons together by means of a continuous exchange of a third particle that moves backwards and forwards between them. We can relate this to everyday experience by imagining two boys representing a proton and a neutron; they remain close to each other because they are playing a game which involves throwing a ball backwards and forwards to each other, Figure 15.7.

The particle is the π-meson and the mass is about 275 times that of an electron. The meson can exist independently for only very short periods (about 10 nanoseconds).

Mass defect

Although we are not certain of the exact nature of the binding force, we can very accurately determine the binding energy of a nucleus. It is found that

Figure 15.7

when protons and neutrons are packed in a nucleus, the mass of the assembled nucleus is less than the sum of the masses of the individual particles. In his theory of special relativity, Einstein showed that mass and energy were equivalent and connected by the equation

$E = mc^2$ where E = energy in joules

m = mass in kg

c = velocity of light in $m\,s^{-1}$

The deficit in mass which occurs when nucleons are packed together into a nucleus is equivalent to the energy needed to bind them together. This deficit is known as the **mass defect**.

   The masses involved in nuclei are very small, so a much smaller unit than the kilogram is used. This unit is the average mass of a particle in the nucleus of a carbon-12 atom. Since this nucleus contains 12 particles and has a mass of 1.992×10^{-26} kg, the mass of this unit (known as the unified atomic mass unit, u) is as follows.

$$1\,u = \frac{1.992 \times 10^{-26}}{12}$$
$$= 1.660 \times 10^{-27}\,kg.$$

Example 1

Find the binding energy of a helium nuclide $^4_2$He if the mass is 4.0015 u.
The helium nuclide $^4_2$He contains 2 protons and 2 neutrons

Individually 2 protons have a mass of $2 \times 1.0078 = 2.0156$ u

2 neutrons have a mass of $2 \times 1.0087 = 2.0174$ u

the total mass of the nucleons is therefore 4.0330 u

However the mass of a helium nuclide is 4.0015 u

There is therefore a deficit in mass, called a mass defect, which is equivalent to the binding energy.

This mass defect $= 0.0315$ u

$= 0.0315 \times 1.660 \times 10^{-27}$

$= 5.23 \times 10^{-29}$ kg

Using Einstein's equation $E = mc^2$

$\Rightarrow$ energy $= 5.23 \times 10^{-29} \times (3 \times 10^8)^2$

$= 4.71 \times 10^{-12}$

The binding energy is 4.71×10^{-12} joules

The binding energy of a nucleon is very small so that it is common practice to state its value in terms of the electron-volt (eV). This is the energy required to move 1 electron through a potential difference of 1 volt.

energy = charge $\times$ p.d.

$\Rightarrow$ $1\,eV = 1.6 \times 10^{-19} \times 1$

$= 1.6 \times 10^{-19}$ J

The larger unit the mega-electron-volt (MeV) which is 1 million electron-volts is normally used.

$1\,MeV = 1.6 \times 10^{-19} \times 10^6$

$\Rightarrow$ $1\,MeV = 1.6 \times 10^{-13}$ J

If Einstein's equation $E = mc^2$ is applied to the unified atomic mass u and the energy converted to MeV, it is found that

$$1\,\text{u is equivalent to 931 MeV of energy.}$$

This conversion will be used in the remainder of this chapter.

15.4   Binding energy per nucleon

When the nucleons in a nucleus are bound together, some of the mass is changed into binding energy. The exact amount of mass which is converted into binding energy varies with the mass number of the element involved.

It is useful to calculate average binding energy per nucleon where

$$\text{average binding energy per nucleon} = \frac{\text{total binding energy of nucleus}}{\text{number of nucleons in nucleus}}$$

A graph of average binding energy per nucleon plotted against mass number (Figure 15.8) provides useful information about the stability of a particular nucleus.

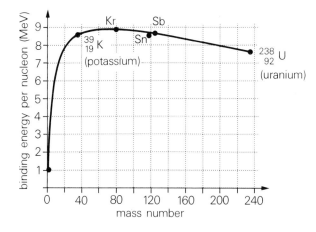

Figure 15.8

It can be seen that the nuclei at each end of the graph, (i.e. those with small and large mass numbers), have smaller average binding energies per nucleon than nuclei with medium mass numbers. Those elements with the highest average binding energy per nucleon are the most stable nuclei.

To see how the graph is constructed consider the following examples.

Example 2

Find the average binding energy per nucleon for uranium-238. ($^{238}_{92}\text{U}$)
(proton mass = 1.0073 u; neutron mass = 1.0087 u;
mass of uranium-238 nucleus = 238.0003 u)

$$\text{mass of 92 protons} = 1.0073 \times 92 = 92.6716\,\text{u}$$

$$\text{mass of 146 neutrons} = 1.0087 \times 146 = 147.2702\,\text{u}$$

$$\text{total mass of individual particles} = 239.9418\,\text{u}$$

$$\Rightarrow \qquad \text{mass defect} = 239.9418 - 238.0003 = 1.9415\,\text{u}$$

but 1 u releases 931 MeV of energy.

$$\Rightarrow \qquad \text{binding energy} = 1.9415 \times 931 = 1807.5365 \text{ MeV}$$

and since there are 238 nucleons present

$$\text{binding energy per nucleon} \qquad = \frac{1807.5365}{238} = 7.6$$

The average binding energy per nucleon is 7.6 MeV

Example 3

Find the average binding energy per nucleon for potassium-39. ($^{39}_{19}$K)
(proton mass = 1.0073 u; neutron mass = 1.0087 u;
mass of potassium-39 = 38.9533 u)

$$\text{mass of 19 protons} = 1.0073 \times 19 = 19.1390 \text{ u}$$
$$\text{mass of 20 neutrons} = 1.0087 \times 20 = 20.1740 \text{ u}$$
$$\text{total mass of individual particles} = 39.3130 \text{ u}$$
$$\Rightarrow \qquad \text{mass defect} = 39.3130 - 38.9533 = 0.3597 \text{ u}$$

but 1 u releases 931 MeV of energy

$$\Rightarrow \qquad \text{binding energy} = 0.3597 \times 931 = 334.8807 \text{ MeV}$$

and since there are 39 nucleons present

$$\text{binding energy per nucleon} = \frac{334.8807}{39} = 8.6$$

The average binding energy per nucleon is 8.6 MeV

15.5   Unstable nuclei

When the number of protons in a nucleus is small, an equal number of neutrons results in a stable nucleus. For example helium has 2 protons and 2 neutrons; oxygen has 8 protons and 8 neutrons. As the number of protons increases, a greater proportion of neutrons is required to ensure stability. This can be seen with iron (26 protons, 28 neutrons), tin (50 protons, 62 neutrons), lead (82 protons, 126 neutrons), and bismuth (83 protons, 126 neutrons). In fact, no nucleus with 84 protons or more is stable. These massive nuclei disintegrate producing radiation.

The disintegration of uranium results in the following radioactive series. In the series each element is formed when the preceding one disintegrates.

element	isotope	radiation		element	isotope	radiation	
uranium	$^{238}_{92}$U	α	γ	lead	$^{214}_{82}$Pb	β	γ
thorium	$^{234}_{90}$Th	β	γ	bismuth	$^{214}_{83}$Bi	β	γ
protactinium	$^{234}_{91}$Pa	β	γ	polonium	$^{214}_{84}$Po	α	
uranium	$^{234}_{92}$U	α	γ	lead	$^{210}_{82}$Pb	β	γ
thorium	$^{230}_{90}$Th	α	γ	bismuth	$^{210}_{83}$Bi	β	
radium	$^{226}_{88}$Ra	α	γ	polonium	$^{214}_{84}$Po	α	
radon	$^{222}_{86}$Rn	α		lead	$^{210}_{82}$Pb	STABLE	
polonium	$^{218}_{84}$Po	α					

The disintegration of the nuclei in such a series will continue until a stable nuclide (usually lead or bismuth) is produced and no more disintegrations take place. Although nuclei with more than 83 protons are unstable, they do exist because all of the nuclei do not disintegrate at the same time.

15.6   Nuclear fission

When a nucleus disintegrates forming a new element, energy is released. This process is known as **nuclear fission**.

The amount of energy released can be found by applying Eihstein's equation of mass-energy equivalence.

For example, uranium-238 disintegrates forming thorium-234 with the emission of an alpha particle.

$$^{238}_{92}\text{U} \longrightarrow \ ^{234}_{90}\text{Th} \ + \ ^{4}_{2}\text{He (alpha particle)}$$
$$238.000 \rightarrow 233.994 + 4.003$$

$$238.000 \rightarrow 237.997 + \text{energy}$$

The total after the disintegration is $237.997\,\text{u}$ so that there is a mass defect of $3 \times 10^{-3}\,\text{u}$. Since $1\,\text{u}$ is equivalent to $931\,\text{MeV}$ of energy, this represents an energy release of $3 \times 10^{-3} \times 931$ which is $2.8\,\text{MeV}$. This type of disintegration is one example of a fission reaction and the energy released is relatively small. If, however, the nucleus is bombarded by a neutron, it splits into two large nuclei and a large amount of energy is released. This was first achieved in 1938 by two Germans Hahn and Strassman who observed the following fission reaction

$$^{238}_{92}\text{U} + ^{1}_{0}\text{n} \rightarrow \ ^{145}_{56}\text{Ba} + ^{94}_{36}\text{Kr}$$

A number of different reactions are possible but in each case two new elements are produced with mass numbers in the range 90 to 150. Some neutrons are normally produced with the release of energy, Figure 15.9.

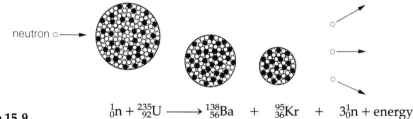

neutron

$$^{1}_{0}\text{n} + ^{235}_{92}\text{U} \longrightarrow \ ^{138}_{56}\text{Ba} \ + \ ^{95}_{36}\text{Kr} \ + \ 3^{1}_{0}\text{n} + \text{energy}$$

Figure 15.9

To describe how the nucleus splits in this way George Gamow, a Russian working in America, proposed a liquid drop model, Figure 15.10.

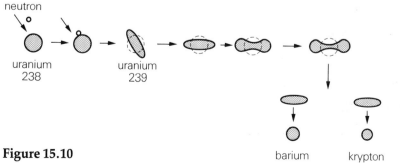

neutron

uranium 238

uranium 239

barium   krypton

Figure 15.10

Firstly the neutron penetrates the 'nuclear liquid' to form the isotope uranium-239. The extra energy causes the nucleus to oscillate. If the energy is sufficient the drop becomes elongated and eventually splits into two separate parts.

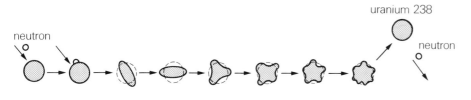

Figure 15.11

If the energy is not sufficient the nucleus vibrates but eventually settles down again releasing a neutron and returning to its original form, Figure 15.11.

To estimate the energy released in such a reaction consider the formation of the stable elements molybdenum and xenon

$$^{235}_{92}U + ^1_0n \longrightarrow ^{98}_{42}Mo + ^{136}_{54}Xe + 2^1_0n + 4^0_{-1}\beta$$

before		after	
$^{235}_{92}U$	234.993 u	$^{98}_{42}Mo$	97.883 u
^1_0n	1.009 u	$^{136}_{54}Xe$	135.878 u
		2^1_0n	2.018 u
		$4^0_{-1}\beta$	0.002 u

total mass 236.002 u   total mass 235.781 u

mass defect   = 0.221 u

energy release = 0.221 × 931

    = 206

The energy released is 206 MeV

Figure 15.12

Uranium-235 is suited to this capture of a neutron and the production of further neutrons (sometimes two and sometimes three) whereas the isotope uranium-238 does not undergo fission easily.

The naturally occurring uranium isotope, uranium-235, comprises only 0.7% of the total amount of natural uranium so that a large build-up of energy produced by a large number of nuclei splitting up at the same time, called a **chain reaction**, is not possible.

However, if the isotope uranium-235 is concentrated in one sample, a process called enrichment, the production of a huge amount of energy is possible. This was first achieved in 1945, on July 16, when the first atomic bomb was exploded at Alamogordo in New Mexico, Figure 15.12.

When the nucleus of uranium splits up, it releases further neutrons which are capable of bombarding neighbouring nuclei producing further fission. If there are sufficient nuclei of uranium-235 available there is a build up of nuclei being split until there is a large release of energy all at once. In order for this to happen a critical mass must be present (about 1 kg) and when this happens the chain reaction goes ahead and an atomic explosion takes place releasing a huge amount of energy, equivalent to thousands of tonnes of conventional explosive. This is shown diagrammatically in Figure 15.13.

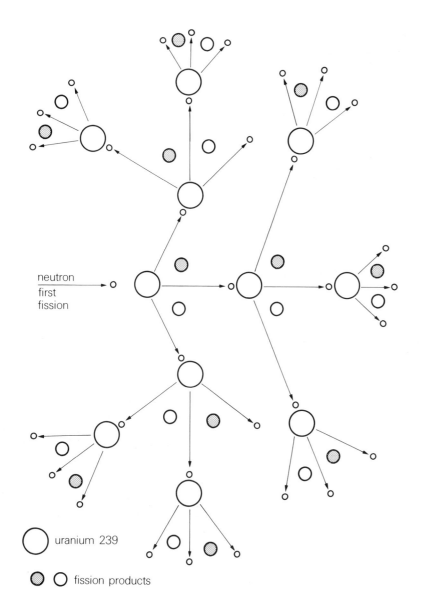

neutron
first
fission

uranium 239

fission products

Figure 15.13

When an atomic explosion takes place there are three damaging effects.

Blast – the explosion forces air outward in huge shock waves and the earth trembles. This combination of wind and earthquake has a devastating effect on houses and structures

Heat – a huge release of heat produces very high temperatures and sets up widespread fires

Radiation – very energetic rays, gamma rays and X-rays, are released which can damage living tissue.

The products of the explosion which are highly radioactive are released and carried by wind and air currents and spread over a large area. These products are known as fallout and are highly dangerous. One of the most dangerous is strontium-90 which does not occur naturally but can be absorbed into the body in much the same way as calcium. This radioactive material persists throughout the lifetime of a human being and increases the risk of cancer.

15.7 Nuclear reactors

In an atomic explosion a chain reaction occurs and a huge uncontrollable release of energy takes place. Fortunately the fission reaction can be controlled so that it produces useful heat energy without the explosion.

For a fission process to continue the following conditions must be satisfied.

1 There are enough heavy nuclei of the fuel material packed together to capture released neutrons. There must be some neutron-producing processes to start the reaction.

2 These neutrons must have the correct energy to cause fission in other nuclei before they escape from the material.

There are three essential parts in a reactor.
a) the rods of fuel which provide heavy nuclei and neutrons
b) control rods which absorb some of the neutrons and so regulate the number of neutrons in the system
c) a moderator which slows down the neutrons.

The first type of operating reactor used natural uranium as the fuel, in the form of rods, with rods of cadmium or boron fitted to control the number of neutrons. Graphite was used as the moderator, all three parts forming the **core** of the reactor.

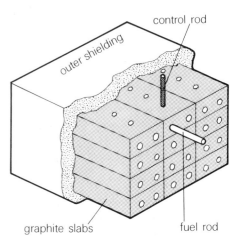

Figure 15.14 Reactor core

The fuel, containing the isotope uranium-235, captures the neutrons and releases a large amount of energy. In order for this to happen, the fast neutrons which are released during fission must be slowed down to increase the chance of capture by the nuclei. Graphite does not capture neutrons but successive collisions with the graphite nuclei result in loss of energy which slows the neutrons down.

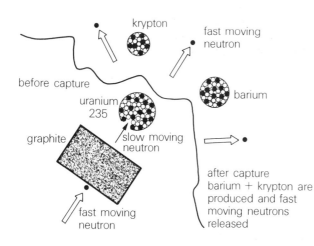

Figure 15.15

The speed of the neutrons must be correct to ensure efficient capturing but another important factor is the neutrons which must be kept at a particular number. This is achieved by using control rods containing a material, such as boron which has a strong affinity for neutrons, to absorb some of the neutrons. Boron captures neutrons forming the stable elements helium ($^4_2$He) and lithium ($^7_3$Li) so that the captured neutron is lost forever. In practice the number of neutrons produced in the reactor is monitored continuously and when it rises above a pre-determined level the control rods are pushed a little further into the core. This additional absorption of neutrons causes the reaction to proceed at a lower rate. If the rods are withdrawn slightly the reaction proceeds at a higher rate.

The nuclear reactor produces heat. This heat raises the temperature of a liquid or gas which circulates round the reactor core. The hot liquid or gas is then used to produce steam which operates a turbine generating electricity, Figure 15.16.

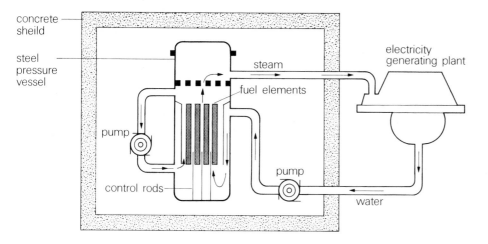

Figure 15.16 Boiling water reactor (BWR)

15.8   Fast breeder reactor

The original reactors used natural uranium (0·7% uranium-235) or slightly enriched fuel (3% uranium-235) but the next set of reactors may use plutonium as the fuel. This is a man-made element produced as a by-product in the original reactors. The process takes place because of the action of neutrons on uranium-238.

$$^{238}_{92}\text{U} + ^{1}_{0}\text{n} \rightarrow ^{239}_{92}\text{U} + 2^{1}_{0}\text{n}$$

followed by

$$^{239}_{92}\text{U} \rightarrow ^{239}_{93}\text{Np} + ^{0}_{-1}\text{e}(\beta)$$
$$\text{neptunium}$$

followed by

$$^{239}_{93}\text{Np} \rightarrow ^{239}_{94}\text{Pu} + ^{0}_{-1}\text{e}(\beta)$$
$$\text{plutonium}$$

The plutonium decays to uranium-235 and the half life of plutonium is very long, about 24 000 years.

$$^{239}_{94}\text{Pu} \rightarrow ^{235}_{92}\text{U} + ^{4}_{2}\text{He}$$

Plutonium undergoes fission in a similar manner to uranium but captures fast neutrons and hence requires no moderator, only control rods to control the rate of the reaction. A typical reaction forms tellurium and molybdenum.

$$^{239}_{94}\text{Pu} + ^{1}_{0}\text{n} \rightarrow ^{137}_{52}\text{Te} + ^{100}_{42}\text{Mo} + 3^{1}_{0}\text{n}$$

Plutonium has been produced in large quantities as a by-product of some reactors and when used in fast breeder reactors (so called because fast neutrons are used), it is possible to convert natural uranium into additional plutonium which can be used as fuel. It thus 'breeds' additional fuel for use in reactors. The fast breeder reactor holds out the possibility of producing large amounts of nuclear fuel. From the known resources of uranium it is estimated that fast breeder reactors could produce the equivalent of 400 years of coal supplies.

Figure 15.17 Fast breeder reactor at Dounreay

The core of the Dounreay prototype reactor

Radioactive materials can cause cancer and produce genetic damage so they must be safely stored in an isolated place. The storage and disposal of the waste from power stations is a major problem and has not yet been adequately solved.

15.9 Nuclear fusion

We have seen that when a heavy nucleus such as uranium splits up forming two new elements, energy is released either in an uncontrolled manner in the atomic bomb or in a carefully controlled process in a nuclear reactor.

It is also possible to release energy by joining two light nuclei together to form a new nucleus in a process known as nuclear fusion.

In order to explain how energy is released by fusion, the average mass per nucleon, measured in unified atomic mass units, must be considered. This is found by dividing the mass of the nucleus by the number of nucleons in the nucleus.

$$\text{average mass per nucleon} = \frac{\text{total mass of nucleus}}{\text{number of nucleons in the nucleus}}$$

Table 1 shows the calculations for some elements.

element	mass of nucleus (u)	No. of nucleons	average mass per nucleon (u)
hydrogen	1.008	1	1.0080
helium	4.002	4	1.0005
potassium	38.953	39	0.9988
tin	119.875	120	0.9990
platinum	193.920	195	0.9945
uranium	234.993	235	1.0000

Table 1

If similar calculations are carried out for all elements, a graph of average mass per nucleon against number of nucleons can be drawn, Figure 15.18.

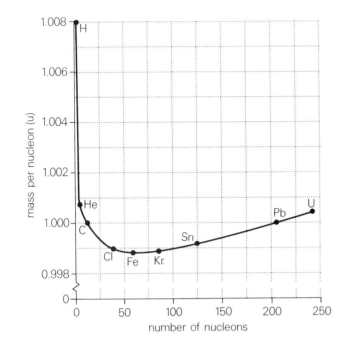

Figure 15.18

Figure 15.19

From the graph it can be seen that if two nuclei of hydrogen are combined to form a helium nucleus, the average mass per nucleon of the single helium nucleus is less than the average mass per nucleon for the original hydrogen nuclei. This deficit in mass is converted into energy.

A number of possible fusion reactions exist and many involve two isotopes of hydrogen, deuterium and tritium, Figure 15.19. Deuterium, usually called heavy hydrogen, contains an additional neutron and can be written either ^2_1H or ^2_1D. It is in plentiful supply existing in 'heavy water' which makes up about 1 part in 5000 of all water. Another isotope, tritium, contains two additional neutrons and can be written ^3_1H or ^3_1T.

In one possible reaction, two nuclei of deuterium combine to form a single nucleus of an isotope of helium plus one neutron.

$$^2_1D \ + \ ^2_1D \ \rightarrow \ ^3_2He \ + \ ^1_0n$$
$$2.013\,u + 2.013\,u \quad 3.015\,u + 1.009\,u$$

There is a mass defect which releases energy

mass defect $= 4.026 - 4.024$
$\qquad\qquad = 0.002\,u$

But $1\,u$ releases about $931\,MeV$

Therefore the energy released $= 0.002 \times 931$
$\qquad\qquad\qquad\qquad\quad = 1.9\,MeV$

A fusion reaction requires very high temperatures in order to provide enough kinetic energy to overcome the forces of electrostatic repulsion between the positive nuclei. Temperatures of one hundred million degrees Celsius are necessary to fuse a large number of nuclei together at the same time so that a thermonuclear explosion can take place: this is the principle of the hydrogen bomb. In the H-bomb the high temperatures are provided by an atomic explosion which is used to initiate the reaction. The high temperatures produce a stream of positive ions and electrons called a plasma.

One possible reaction which can take place in the plasma is described by the following equation

$$^2_1H + ^3_1T \rightarrow ^4_2He + ^1_0n + 17.6\,MeV$$

The energy released, 17.6 MeV, is very much less than that obtained by a fission reaction (200 MeV) but it must be remembered that the nucleus of uranium has a mass of about 235 u whereas the mass of six deuterium nuclei is about 12 u. The energy yield per kilogram for an H-bomb explosion is very much larger than the energy per kilogram for an A-bomb explosion. There is also no requirement for a critical mass as there is for the A-bomb.

The possibility of producing energy from a controlled thermonuclear reaction is very attractive because of the plentiful supply of deuterium and also because the radioactive waste produced is small.

In order to achieve a controlled thermonuclear reaction four conditions must be satisfied.
a) suitable material which will undergo fusion must be available
b) material must be heated to the required high temperature
c) hot plasma must be contained sufficiently long to allow the energy produced by fusion to exceed the input energy
d) fusion energy must be converted into electricity

A possible practical arrangement is illustrated in Figure 15.20. This shows the Tokamak system first developed in the USSR in the late 1960's.

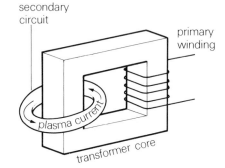

Figure 15.20

A current is produced in the primary circuit, usually by some kind of discharge. Transformer action causes a very large current to be induced in the secondary. This current heats the plasma and temperatures of 70 million degrees have been attained.

The most difficult problem is to contain the plasma because if the plasma touches the walls of the container the high temperatures would destroy the container material. Present methods use strong magnetic fields which are designed to confine the stream of plasma. The magnetic field produced by the plasma current itself tends to hold the particles of the plasma together. The streams of ions moving in the same direction produce magnetic fields which tend to draw the streams together in the same way that two conductors lying side by side will be pulled towards each other by the magnetic fields (Figure 15.21). This is known as the **pinch effect** (Figure 15.22).

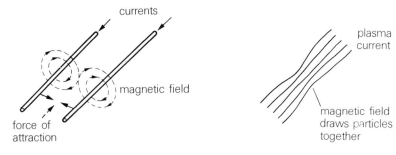

Figure 15.21   **Figure 15.22**

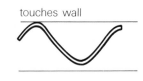

Figure 15.23

Unfortunately the plasma does not remain stable and starts to twist and eventually touches the walls, Figure 15.23.
The plasma can be stabilized for short periods by applying a large magnetic field by means of a coil wrapped round the container walls, Figure 15.24.
This and other methods are not entirely satisfactory and a great deal of further research is needed to produce a working system. The first attempt at a

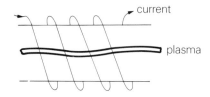

Figure 15.24

thermonuclear reactor in the United Kingdom was completed at Harwell in 1958, Figure 15.25. It was known as ZETA.

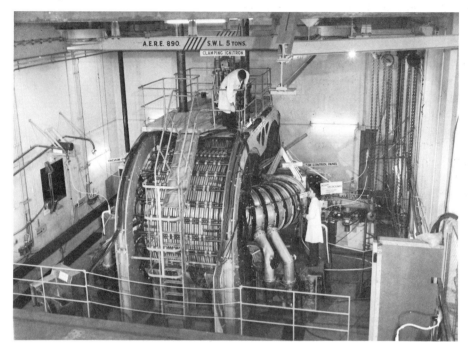

Figure 15.25

A circular aluminium tube, known as a torus, contains deuterium. The tube has a diameter of about 1 metre and the circle is about 4 metres across, Figure 15.26.

To start up the process a radio frequency oscillator is used to produce a spark discharge inside the torus which ionizes the deuterium gas. This oscillator is turned off and a large bank of capacitors is discharged through the primary winding of a transformer. This causes a large pulse of deuterium ions to flow round the torus which forms the secondary winding of the transformer. Currents of 250 000 amperes existing for 2 milliseconds have been recorded.

However the plasma cannot be easily stabilized in the circular-shaped torus and research is at present progressing using different methods. Eventually a commercially successful thermonuclear reactor may be built but it will not provide cheap power. There are formidable engineering difficulties and the capital cost of such a reactor is high because of the equipment needed for magnets, cooling systems and ancillary equipment. As supplies of conventional fuel begin to run out the cost of a thermonuclear reactor will become competitive. The exciting aspect of the system is the fact that the deuterium fuel is in such plentiful supply in the oceans of the world.

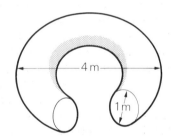

Figure 15.26

Summary

Rutherford's model of the atom assumes that the mass and positive charge are concentrated in the nucleus surrounded by a space containing the negative charge. The volume of the nucleus is extremely small compared with the total volume of the atom.

The nucleus of an atom contains protons and neutrons.

A nucleus can be described using a symbol such as $_Z^A X$ where X denotes the element and

Z = number of protons
$A - Z$ = number of neutrons

Isotopes contain the same number of protons but different numbers of neutrons.

An element can have more than one isotope.

1 unified atomic mass unit $1\,u = 1.66 \times 10^{-27}\,kg$.

1 electron-volt is equal to $1.6 \times 10^{-19}\,J$.

Einstein showed that mass and energy are related by the equation $E = mc^2$.

In nuclear fission a nucleus splits forming two new elements and releasing energy.

A critical mass is required before a chain reaction can take place.

In nuclear fusion two nuclei join together forming a single nucleus and releasing energy.

Problems

1   Two assistants of Rutherford, Geiger and Marsden, conducted an experiment in order to investigate the nature of the atom. Draw a diagram showing the experimental arrangements used and describe how they collected data from the experiment. What model of the atom did this experiment confirm?

2   Explain the meaning of the following: electron, proton, neutron, isotope.

3   Define mass number and atomic number.

4   Describe a simple form of mass spectrometer.

5   Einstein put forward the relationship $E = mc^2$. Use this equation to calculate the energy released when 1 milligram of mass is converted into energy.

6   Energy is released when a nucleus splits up forming two new nuclei but energy can also be released when two nuclei combine to form one single nucleus. Explain why each of these processes can produce energy.

7   In a nuclear reactor a controlled fission process releases energy. Draw a diagram showing the stages in the reactor leading up to the generation of electricity.

8   A neutron undergoes decay forming a proton. The equation of the reaction is

$$_0^1 n \rightarrow {}_1^1 H + {}_{-1}^0 e$$

Determine the energy released during this reaction.
(mass of $_0^1 n = 1.0087\,u$; mass of $_1^1 H = 1.0078\,u$; mass of $_{-1}^0 e = 0.0006\,u$)

9   Calculate the binding energy in MeV of the $_3^7 Li$ nucleus.
(mass of nucleus = $7.0144\,u$; mass of a proton = $1.0078\,u$; mass of $_0^1 n = 1.0087$)

10   Two deuterons fuse together to form a helium nucleus. The equation of this reaction is

$$_1^2 H + {}_1^2 H \rightarrow {}_2^3 He + {}_0^1 n$$

Calculate the energy released. (mass of $_1^2 H = 2.0136\,u$; mass of $_2^3 He = 3.0149\,u$; mass of $_0^1 n = 1.0087\,u$

11   Determine the binding energy of a helium $_2^4 He$ nucleus if the mass of the nucleus is $4.0015\,u$.

12   How much energy will be released when 0.5 kg of uranium is completely transformed into energy?

13   Explain briefly the function of a moderator in a fission reactor.

14   In a fission reaction the average energy released is 175 MeV per fission. How many such fissions are required per second in order to provide a power of 1 MW?

15   Plutonium undergoes fission producing tellurium and molybdenum.

$$_{94}^{239} Pu + {}_0^1 n \rightarrow {}_{52}^{137} Te + {}_{42}^{100} Mo + 3\,{}_0^1 n$$

Calculate the energy released from this reaction. The mass of $_{52}^{137} Te$ is not accurately known but can be taken as $137.0000\,u$.
(mass of $_{94}^{239} Pu = 239.0006\,u$; mass of $_{42}^{100} Mo = 99.8850\,u$)

16   The carbon-12 nucleus $^{12}_6$C has a mass of 11.9967 u.
Calculate the binding energy of the nucleus. What is the binding energy per nucleon?

17   Explain why some types of reactor do not require a moderator.

18   What is spontaneous fission? What isotope is known to be spontaneously fissionable?

19   Why is the controlled fusion of hydrogen into helium such a difficult problem?

20   Uranium splits up forming molybdenum and xenon which are stable. The equation describing this reaction is

$$^{235}_{92}U + {}^1_0n \rightarrow {}^{98}_{42}Mo + {}^{136}_{54}Xe + 2{}^1_0n$$

Calculate the energy released. (mass of $^{98}_{42}$Mo = 97.8830 u; mass of $^{136}_{54}$Xe = 135.8776 u; mass of $^{235}_{92}$U = 234.9934 u)

21   In 1932 Cockroft and Walton produced nuclear disintegrations by accelerating protons with a high voltage machine.
The reaction can be written

$$^7_3Li + {}^1_1H \rightarrow {}^4_2He + {}^4_2He$$

Calculate the energy released. (mass of $^7_3$Li = 7.0144 u; mass of $^4_2$He = 4.0015 u; mass of $^1_1$H = 1.0078 u)

22   Explain the function of the control rods and the moderator in a nuclear reactor.

23   State the conditions necessary for a chain reaction to take place.

24   A proton and a neutron combine to form deuterium $^2_1$H.

$$^1_1H + {}^1_0n \rightarrow {}^2_1H$$

Calculate the binding energy of deuterium. (mass of $^2_1$H = 2.0136 u)

25   Calculate the total energy released during the following reaction.

$$^{238}_{92}U \rightarrow {}^{234}_{90}Th + {}^4_2He$$

(mass of $^{238}_{92}$U = 238.0003 u; mass of $^{234}_{90}$Th = 233.9942 u; mass of $^4_2$He = 4.0015 u)

26   In a nuclear power station, liquid sodium is used to remove heat from the core of the reactor which is at a temperature of 833 K. This process is carried out in two stages. Hot sodium circulating in the core transfers heat to more molten sodium in a secondary system at the first heat exchanger. Heat from the sodium in the secondary system is then used to produce steam at the second heat exchanger. The steam is used to drive the turbine-generators and so produce electricity.

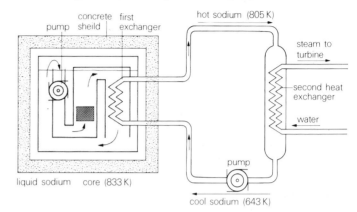

a)   Molten sodium flows through the second heat exchanger at a rate of 1.74×10^5 kg every minute, entering the exchanger at a temperature of 805 K and leaving at 643 K. Calculate the rate, in watts, at which energy is transferred to the water system. You may use any information from the table in part (c).

b)   How will the electrical power output from the generators compare with the value you have calculated in part (a)? Explain your answer.

c)   The table below lists some properties of a number of materials.

Material	Melting point K	Boiling point K	Specific heat capacity of liquid J kg^{-1} K^{-1}	Heat conductivity of liquid
sodium	371	1156	1280	good
aluminium	932	2720	1085	good
water	273	373	4200	poor
lead	601	2010	139	good

By comparing the properties of sodium with those of each of the other materials in the table, discuss why sodium is chosen for use in the reactor in preference to any of the others.

d)   Suggest why the sodium circulating in the core of the reactor is not used to carry energy directly to the second heat exchanger.

SCEEB

Numerical answers

Chapter 1

2. $2.5\,m\,s^{-1}$ 53·1° W of N
3. $3.81\,m$, 23° above the horizontal
4. a) $30\,m\,s^{-1}$ b) $8.33 \times 10^7\,m\,s^{-1}$
   c) $6.38\,m\,s^{-1}$
5. a) $0.12\,m\,s^{-1}$, $0.84\,m\,s^{-1}$, $1.92\,m\,s^{-2}$
   b) $0.48\,m\,s^{-1}$
6. a) $1.5\,m\,s^{-2}$ b) $1.08\,m$
8. a) $3\,s$ b) $22.5\,m$
9. $0.47\,m\,s^{-1}$
12. $1.13\,m$
13. $99.9\,m$
14. zero
16. a) $1\,m\,s^{-1}$ b) $6.5\,m$
17. ′ a) zero b) $6\,m\,s^{-2}$ c) $-2\,m\,s^{-2}$
18. $237.5\,m$
19. a) $22\,m\,s^{-1}$ downwards b) $21\,m$
22. a) $0.5\,s$ b) $5\,m\,s^{-1}$ downwards
    c) 64·4° below the horizontal
    d) $1.2\,m$
23. a) $0.5\,s$ b) $2.165\,m$
24. a) $7.2\,m$ b) $14.4\,m$
25. a) $17.3\,m\,s^{-1}$
    b) $5\,m$ above the top of the launcher
26. a) $140\,m$ b) $5\,m$
    c) $40.3\,m\,s^{-1}$ 29·7° above the horizontal
28. a) $6.3\,m\,s^{-1}$
    b) $0\,m, 2\,m, 8\,m, 16\,m, 24\,m$
    c) (i) $45.5\,m$ from the start.
29. b) $7.5\,m\,s^{-2}$ c) (ii) $0.55\,s$
    (iii) $9.52\,m\,s^{-2}$
30. b) $25\,m\,s^{-1}$, 53° below the horizontal.

Chapter 2

2. a) $510\,kN$ b) $858\,kN$
3. a) $0.293\,J$ b) $0.71\,N$ c) $7.1\,m\,s^{-2}$
4. a) $25\,N$ b) $6\,m$
7. a) $22.4\,m\,s^{-1}$ b) $0.12\,N$
9. $0.5\,m\,s^{-1}$ $0.375\,J$ $1.125\,J$
10. $0.8\,s$
12. a) $36\,N\,s$ b) $36\,kg\,m\,s^{-1}$ c) $432\,J$
13. a) $287.5\,m$ b) $3.69\,kN$
14. b) $3500\,m$
15. a) $0.75\,m\,s^{-1}$ b) $0.03\,m$
16. 25.1 milliseconds
17. a) $0.4\,m\,s^{-2}$, $0.2\,m\,s^{-2}$,
    zero, $-0.6\,m\,s^{-2}$
    b) $70\,kN$
    c) $360\,kW$
18. $3\,J$
19. a) $12\,J$ b) $1.6\,J$ c) $5.2\,N$
20. a) $2\,m\,s^{-1}$ c) $36\,m$
21. a) $0.032\,N\,s$ b) $0.71\,N$
    c) $0.0064\,J$

Chapter 3

1. $0.64\,C$
2. $5 \times 10^3\,A$
3. $2 \times 10^{-3}\,s$
4. $6\,V$
6. 0.5 ohms
7. $37.5\,s$
8. $1\,A$
9. $0.8\,V$
10. $5\,A$
11. 6 ohms, $27\,V$
12. $0.1\,A$
13. $6\,V$, 1.2 ohms, $5\,A$
14. $1.5\,V$, 0.05 ohms, $30\,A$
15. $3.33\,W$
16. a) $R = r$ b) $1.44\,kW$
17. $R/(R+r)$
18. b) (ii) $5\,V$, 2.5 ohms
19. a) (i) $9\,W$ (ii) $3\,V$
    b) (ii) 3 ohms (iii) 6 ohms

Chapter 4

1. a) $0.8\,V, 1.6\,V, 3.2\,V, 6.4\,V$
   b) $0.8\,V, 1.6\,V, 2.4\,V, 3.2\,V, 4.8\,V, 5.6\,V,$
   $6.4\,V, 9.6\,V, 11.2\,V, 12.0\,V.$
2. $0-3\,V$
4. 4.1 ohms
6. $0.62\,A$
7. $2\,V$ less
9. 2 ohms
11. 121.5 ohms
13. 30.5 ohms
14. 4.1 ohms
15. b) 20 ohms, 2 ohms
    c) $55\,mA$, $0.99\,V$
16. a) 102 ohms b) $\frac{3}{20}$

Chapter 1 (continued — column 2)

22. b) $140\,N$ downwards, zero,
    $140\,N$ upwards
23. a) $1.0\,N$
24. a) (i) $1.76\,N$ (ii) $0.25\,s$
25. a) $10\,m\,s^{-1}$ b) (1) $2.5\,m\,s^{-1}$
    c) (i) $6250\,J$ (ii) $2400\,J$
    d) $7.2 \times 10^4\,N$
26. a) (i) $0.7\,m\,s^{-2}$
    (ii) $3 \times 10^4\,N$
28. a)

height h (m)	0.60	1.00	1.40	1.80
speed at B ($m\,s^{-1}$)	2.50	3.85	4.90	5.80
potential energy at A (J)	6.0	10.0	14.0	18.0
kinetic energy at B (J)	3.13	7.40	12.0	16.8

   b) $1.30\,N$ c) (ii) $0.34\,m$

Chapter 4 — right column

17. a) 0.01 ohms
    b) 599 ohms in series c) $1\,V$, $0.8\,V$
18. b) $900\overset{+}{-}37$ ohms
19. b) $25\,037$ ohms b) $16\,667$ ohms

Chapter 5

4. $1.92 \times 10^{-16}\,N$
5. $6.7 \times 10^3\,N\,C^{-1}$
6. $4 \times 10^{-8}\,C$
7. $1.8 \times 10^5\,N\,C^{-1}$
8. $3.6 \times 10^3\,V$
9. $8.6 \times 10^{-19}\,C$
10. a) 4 b) 2
11. $1.76 \times 10^{15}\,m\,s^{-2}$

Chapter 6

5. $60\,\mu C$
6. $5 \times 10^{-10}\,F$
8. $5 \times 10^{-4}\,F$
13. a) $0.1\,A$ b) $10\,V$ c) $1 \times 10^{-3}\,C$
14. $3 \times 10^{-2}\,C$, $0.225\,J$
15. $3.3\,\mu F$
16. b) $5000\,\mu F$
17. c) $0-1\,mA$
18. a) $6 \times 10^{-3}\,C$

Chapter 7

9. $2\,A\,s^{-1}$
10. $120\,V$
11. $2.4\,mH$
12. $6\,mA$, $24\,A\,s^{-1}$

Chapter 8

3. $17\,V$
4. $24\,W$
7. 31.8 ohms, 0.63 ohms, 5 ohms
8. $5\,Hz$
9. $2 \times 10^{-8}\,H$
13. $80\,Hz$
15. b) $200\,Hz$

Chapter 9

1. 1.93
2. $2 \times 10^8\,m\,s^{-1}$
5. a) $4.8 \times 10^{14}\,Hz$
   b) $4 \times 10^{14}\,Hz$ c) $4.17 \times 10^{-7}\,m$
   d) $2.0 \times 10^8\,m\,s^{-1}$
7. a) 48·8°
8. b) 0·5°
9. a) 1.47
10. a) 1.51 c) $1.99 \times 10^8\,m\,s^{-1}$
11. $0.125\,W\,m^{-2}$
12. $64\,W\,m^{-2}$ $7.1\,W\,m^{-2}$ $4.0\,W\,m^{-2}$
    $2.6\,W\,m^{-2}$

Chapter 10

2. a) (i) 13 cm   (ii) 13 cm
3. a) (i) 10 cm   (ii) 8 cm
5. a) (i) 17 cm   (ii) 20 cm   (iii) 30 cm
    (iv) −15 cm
6. a) (i) −3·5 cm   (ii) −3·3 cm   (iii) −2·5 cm
    (iv) −2·2 cm

8. a) 5 D   b) 22·5 D   c) 13·3 D
9. a) (ii) 6
  b) (ii) 10 cm to 20 cm
10. a) 0·15 m   b) 0·01 m

7.

lens	object distance	image distance	focal length	image	magnification
		−60 cm		virtual	4
converging			10 cm		1
diverging		−5·5 cm		virtual	0·55
		10 cm	3·3 cm	real	

Chapter 11

9. 3 D

Chapter 12

4. 6×10^{-7} m
5. $3·6 \times 10^{-2}$ m
6. $4·8 \times 10^{-7}$ m
9. 27°
10. b) 9·0°
11. a) 7×10^{-7} m   b) $2·50 \times 10^{-5}$ m
    c) 7×10^{-7} m   d) 4×10^{-7} m
12. a) 0·66 μ   b) 660 nm
    c) $6·6 \times 10^{3}$ Å
15. d) $6·25 \times 10^{-7}$ m

Chapter 13

1. 60 kPa
2. $8·0 \times 10^{8}$ Pa
6. $1·44 \times 10^{-10}$ m³; $9·90 \times 10^{-10}$ m
7. $3·73 \times 10^{-26}$ m³; 0·089 kg m⁻³
10. $3·91 \times 10^{-4}$ m³
11. 0·067 m³
12. 238·3 kPa
13. 400 kPa
14. $8·81 \times 10^{25}$
15. 1·25 kg m⁻³
16. 1600 kg m⁻³; $8·08 \times 10^{-4}$ m³;
17. 265 cm³
18. 5·28 kg
20. 8·29 J K⁻¹
22. a) 34·00 cm³
    b) (i) 21·94 cm³   (ii) 12·06 cm³

23. b) 580 m s⁻¹
24. a) $1·40 \times 10^{5}$ b) 1·03 : 1
26. $1·27 \times 10^{26}$
28. b) (ii) 4338 K
29. b) 23 cm   c) (ii) $1·28 \times 10^{5}$

Chapter 14

1. $0·4 \times 10^{-19}$ C; 0–2500 V
2. a) 2000 V cm⁻¹   b) $3·2 \times 10^{-19}$ C
4. a) $5·0 \times 10^{14}$ Hz   b) (i) no (ii) yes.
5. a) $1·5 \times 10^{-17}$ J   b) $2·2 \times 10^{16}$ Hz
6. 16·6 V
7. a) 6   b) $3·0 \times 10^{15}$ Hz, $5·7 \times 10^{14}$ Hz
10. (ii) $2·9 \times 10^{-19}$ J

Chapter 15

5. 9×10^{10} J
8. 0·28 MeV ($4·5 \times 10^{-14}$ J)
9. 40·8 MeV ($6·6 \times 10^{-12}$ J)
10. 3·4 MeV ($5·4 \times 10^{-13}$ J)
11. 29·4 MeV ($4·7 \times 10^{-12}$ J)
12. $4·5 \times 10^{16}$ J
14. $3·6 \times 10^{16}$ per second
15. 91·4 MeV ($1·46 \times 10^{-11}$ J)
16. 95 MeV ($1·52 \times 10^{-11}$ J),
    7·9 MeV ($1·27 \times 10^{-12}$ J)
20. 208·6 MeV ($3·3 \times 10^{-11}$ J)
21. 17·9 MeV ($2·9 \times 10^{-12}$ J)
24. 2·7 MeV ($4·3 \times 10^{-13}$ J)
25. 4·3 MeV ($6·9 \times 10^{-13}$ J)
26. a) $6·01 \times 10^{8}$ W

Index

Acknowledgements

The publishers and authors would like to thank the following for permission to reproduce photographs:

Barnaby's Picture Library, p.1; T. Bodford-Cousins/The National Meteorological Office, p.54; Paul Brierley, p.136 bottom left and right, p.137, p.138; Peter Dazely, p.9; Griffin and George Ltd., p.88; Hydraulics Research Station, Wallingford, p.136 top (Crown Copyright); The Johns Hopkins University, p.33; The Los Alamos National Laboratory, p.192; Mansell Collection, p.80; D.M. Nicholas, p.150; Photographs and All That, p.50, p.60, p.62, p.65, p.67, p.80 top, p.101, p.146; United Kingdom Atomic Energy Authority, p.196 left and right, p.199; Volvo, p.20.

Notes

Notes

Notes

Notes

Notes

Notes

Notes

Notes